AF774130

Corrosion

Edited by Ambrish Singh

Published in London, United Kingdom

IntechOpen

Supporting open minds since 2005

Corrosion
http://dx.doi.org/10.5772/intechopen.87880
Edited by Ambrish Singh

Contributors
Shikha Munjal, Said Abbout, Ambrish Singh, Kashif Ansari, Mumtaz Ahmad Quraishi, Dheeraj Singh Chauhan, Savas Kaya, Hua Yu, Yuanhua Lin, Peter Hammer, Andressa Trentin, Samarah Vargas Harb, Thiago A. C. De Sousa, Mayara C. Uvida, Ronaldo Cozza, Marcelo Macedo, Vladimir A. Pukhliy, Anil Kumar, Alex Lugovskoy, Svetlana Lugovskoy, Elinor Nahum, Michael Schorr, Benjamin Valdez, Margarita Stoytcheva, Roumen Zlatev, Ekemini Ituen, Onyewuchi Akaranta Onyewuchi Akaranta, Vikas Verma, Jorge Humberto Luna-Domínguez, Aakash Singh

© The Editor(s) and the Author(s) 2020
The rights of the editor(s) and the author(s) have been asserted in accordance with the Copyright, Designs and Patents Act 1988. All rights to the book as a whole are reserved by INTECHOPEN LIMITED. The book as a whole (compilation) cannot be reproduced, distributed or used for commercial or non-commercial purposes without INTECHOPEN LIMITED's written permission. Enquiries concerning the use of the book should be directed to INTECHOPEN LIMITED rights and permissions department (permissions@intechopen.com).
Violations are liable to prosecution under the governing Copyright Law.

CC BY

Individual chapters of this publication are distributed under the terms of the Creative Commons Attribution 3.0 Unported License which permits commercial use, distribution and reproduction of the individual chapters, provided the original author(s) and source publication are appropriately acknowledged. If so indicated, certain images may not be included under the Creative Commons license. In such cases users will need to obtain permission from the license holder to reproduce the material. More details and guidelines concerning content reuse and adaptation can be found at http://www.intechopen.com/copyright-policy.html.

Notice
Statements and opinions expressed in the chapters are these of the individual contributors and not necessarily those of the editors or publisher. No responsibility is accepted for the accuracy of information contained in the published chapters. The publisher assumes no responsibility for any damage or injury to persons or property arising out of the use of any materials, instructions, methods or ideas contained in the book.

First published in London, United Kingdom, 2020 by IntechOpen
IntechOpen is the global imprint of INTECHOPEN LIMITED, registered in England and Wales, registration number: 11086078, 5 Princes Gate Court, London, SW7 2QJ, United Kingdom
Printed in Croatia

British Library Cataloguing-in-Publication Data
A catalogue record for this book is available from the British Library

Additional hard and PDF copies can be obtained from orders@intechopen.com

Corrosion
Edited by Ambrish Singh
p. cm.
Print ISBN 978-1-78985-255-4
Online ISBN 978-1-78985-256-1
eBook (PDF) ISBN 978-1-83962-738-5

For EU product safety concerns:
IN TECH d.o.o., Prolaz Marije Krucifikse Kozulić 3, 51000 Rijeka, Croatia, info@intechopen.com or visit our website at intechopen.com.

We are IntechOpen, the world's leading publisher of Open Access books

Built by scientists, for scientists

5,000+
Open access books available

125,000+
International authors and editors

145M+
Downloads

151
Countries delivered to

Our authors are among the
Top 1%
most cited scientists

12.2%
Contributors from top 500 universities

CLARIVATE ANALYTICS BOOK CITATION INDEX INDEXED

WEB OF SCIENCE™

Selection of our books indexed in the Book Citation Index
in Web of Science™ Core Collection (BKCI)

Interested in publishing with us?
Contact book.department@intechopen.com

Numbers displayed above are based on latest data collected.
For more information visit www.intechopen.com

Meet the editor

Dr. Ambrish Singh is currently working as a Professor in the School of New Energy and Materials, Southwest Petroleum University, China. His research interests are mainly focused on corrosion, electrochemistry, green chemistry, quantum chemistry, smart coatings, polymers, nano-materials, composites, and petroleum engineering. He has received the prestigious Sichuan 1000 Talent Award, the President's Award, and the Young Scientist Award for his outstanding research contributions. He has published more than 100 SCI research papers, acting as reviewer for more than 40 journals and editor for several journals. Dr. Singh has been invited to present his work at several national and international conferences, seminars, and workshops. He has finished several state and provincial projects in China and India. He is an active member of NACE, SPE, and ACS.

Contents

Preface

Corrosion is a big problem globally and needs to be minimized to boost the economy. Corrosion is omnipresent in all metallic systems in all parts of the world. All systems used in engineering are subject to corrosion, which leads to their degradation, including systems used in clean energy, water supply, construction, transportation, food, commercial world, treatment plants, boilers, and storage tanks. Corrosion is part of our life from stents used in the heart, wires used in teeth, Mg/Ti implants in the body, to nuclear power plants. Due to its presence in industries and multidisciplinary areas, corrosion is one of the hot topics of research for young scientists and students who are placing their feet in the shoes of other professors and eminent scientists to find a suitable solution.

Corrosion is a big threat resulting in the loss of billions of dollars each year. This economic meltdown needs to be mitigated through proper and advanced techniques. Scientists and authors have come up with new techniques comprising of ceramics, polymers, glass, and other materials in line with the existing techniques. However, there is an ongoing demand to meet the environment regulations and produce low cost corrosion inhibitors and coatings. This book includes different sections with several chapters embedded in it. The sections include chapters on corrosion inhibitors, biological corrosion, implants, and general corrosion. All these chapters will serve the academicians, researchers, and industrialists to grasp deeper insights into the mechanism and mode of action of corrosion on different metals in various aggressive solutions.

Dr. Ambrish Singh
Professor,
School of New Energy and Materials,
Southwest Petroleum University,
Chengdu City, China

Section 1

Corrosion Inhibitor

Chapter 1

Corrosion Mitigation by Planar Benzimidazole Derivatives

Ambrish Singh, Kashif R. Ansari, Dheeraj S. Chauhan, Mumtaz A. Quraishi, Savas Kaya, Hua Yu and Yuanhua Lin

Abstract

The corrosion has a considerable amount of impact on the economics of every nation, and ultimately it affects the GDP. In the present era, the challenge given by corrosion can be easily mitigated using organic compounds as corrosion inhibitor in different corrosive media. The important property of an inhibitor is the presence of the metal interacting with heteroatoms and a planar structure. In this regard, benzimidazoles (BI) with a fused bicyclic ring consisting of benzene and imidazole moiety in their structural framework making them a potential candidate for anti-corrosion work. In addition to this, bezimidazole derivatives are classified as green inhibitor due to different kinds of biological activities. Their higher potency to mitigate corrosion is because of the planar molecular structure, nitrogen atom and sp2 hybridized carbon, which provide them an ability to strongly interact with the metal. The focus of this book chapter is to investigate briefly the anti-corrosion ability of benzimidazole (BI) and their derivatives as a potential corrosion inhibitor for various industrially useful metals in different aggressive media.

Keywords: benzimidazole, corrosion, inhibitor

1. Introduction

The costs and the aftermath of corrosion are enormous. There are so many reports identified, resulting in the death and injury of living beings due to the corrosion failure. The easily quantifiable costs are those associated with the repair and replacement of equipment. It is very much difficult to predict the exact loss associated with the corrosion especially when corrosion causes the production unit in the petroleum industry to shut down. The potential method to reduce the corrosion is using the organic compounds in the aggressive media for protecting the metals against corrosion. Among the numerous inhibitor groups, benzimidazole derivatives have gained a more practical application due to their environmentally benign nature, potential ability to strongly interact with the metal surface through the nitrogen atoms and easy methods to modify their molecular structure as per the required need.

The benzo derivative of imidazole is referred to as benzimidazole. Although benzimidazole is the commonest name of the parent compound of the series, other names such as benzimidazole and 1,3-benzodiazole are often used. The main purpose of this book chapter is to introduce briefly the anti-corrosion ability of benzimidazole and their derivatives as a potential corrosion inhibitor for the various industrially useful metals in different aggressive media.

1.1 A brief overview on biological activities

The molecular structure of benzimidazole arises by fusing the benzene and imidazole rings. It generates a class of bicyclic compounds which have a wide range of biological and pharmacological activities (**Figure 1**).

The molecular structures of some benzimidazole containing nucleus compounds, which are commonly used for various therapeutic applications, are presented in **Figure 2**: (1) benoxaprofen analog (anti-inflammatory), (2) pantoprazole (proton pump inhibitor), (3) candesartan (antihypertensive), (4) albendazole (anthelmintics/antimicrobial), (5) mebendazole (anthelmintics), (6) bilastine (antihistaminic), (7) carbendazim (antifungal), (8) bendamustine (antitumor) and (9) astemizole (antihistaminic) [1].

1.2 Molecular structure

The general molecular structure consists of a bicyclic heteroatomic structure with two nitrogen atoms (**Figure 3**).

The benzimidazole derivatives have different kinds of extensive coupling constant, which leads to generate coupling with strong force constants [2].

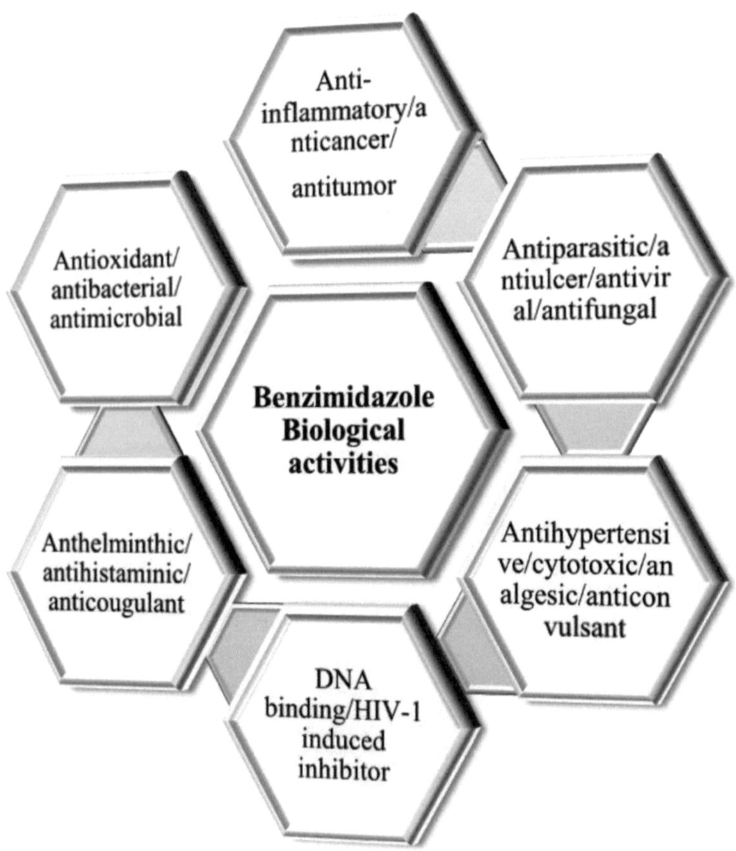

Figure 1.
Biological activities of benzimidazole.

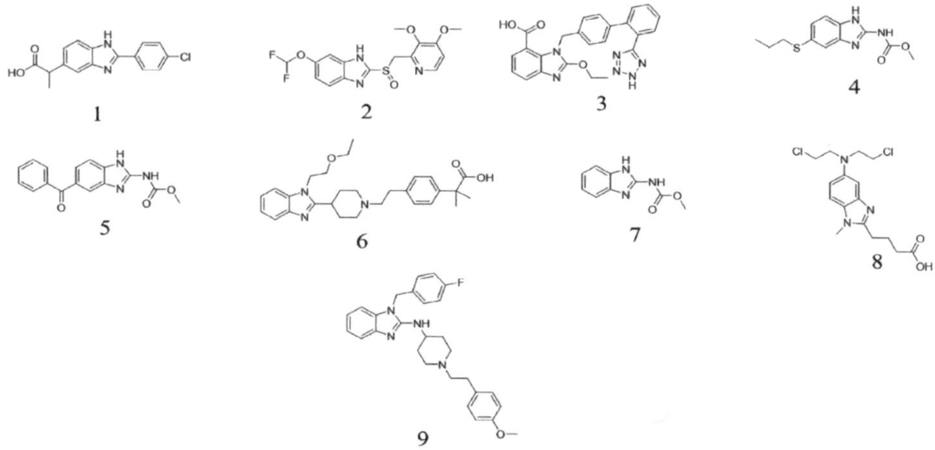

Figure 2.
Therapeutic applications of benzimidazole derivatives.

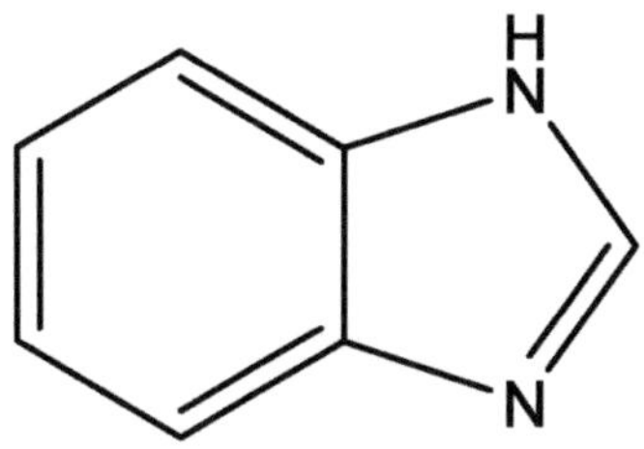

Figure 3.
Molecular structure of benzimidazole.

Benzimidazole derivatives have a greater affinity to produce a variation of chemical species because of the high force constant, and this is a remarkable characteristic of the benzimidazole. The analysis of vibrational property of benzimidazole derivatives is difficult due to the presence of extensive coupling in their molecular structure [3], although scientists were successful to analyse the vibrational properties of simple benzimidazole derivatives by applying density functional theory (DFT).

1.3 Synthesis

The synthesis of benzimidazole is very important with respect to their varying potential pharmacological activities. Thus, a unique attention is required to fulfil this process. In literature, various methods are given using different kinds of catalyst. Few of them are as follows:

Azarifar et al. [4], synthesized benzimidazole derivatives by condensation of o-phenylenediamine promoted by acetic acid under microwave. They concluded that a mild, manipulatable procedure, eco-friendly and green aspects avoiding hazardous solvents, shorter reaction times and high yields of the products are the advantages of this method.

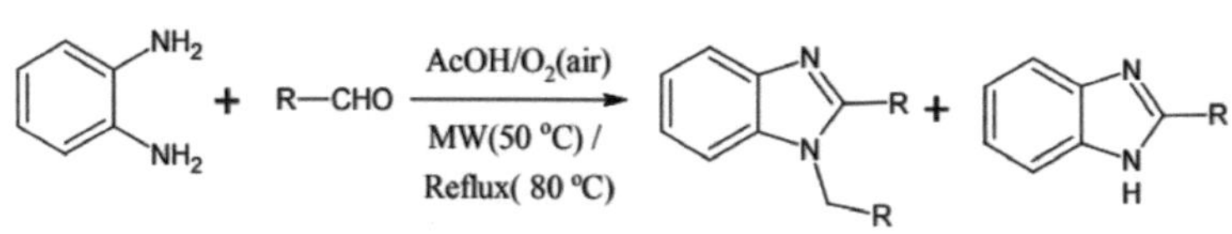

Shaikh et al. [5] have efficiently synthesized benzimidazoles in high yields by treatment of 1,2-diamine with aldehydes using the metal coordinate complex K4[Fe (CN)6] as a catalysis.

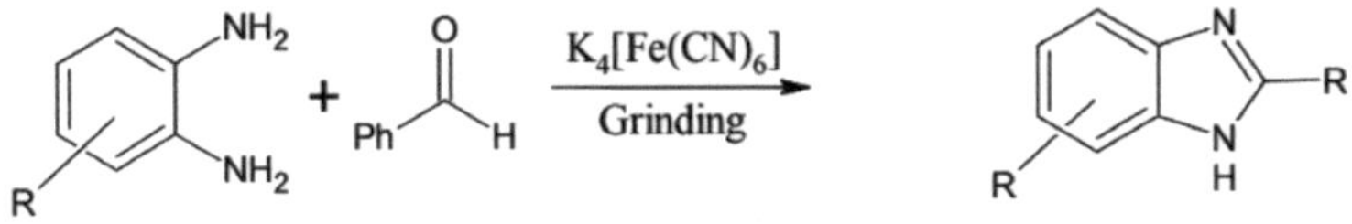

Vaidehi et al. [6] had synthesized a set of 2-substituted benzimidazoles successfully by condensation of o-phenylenediamine with substituted acids in the presence of ring closing agents like polyphosphoric acid/HCl. The present work has demonstrated the use of a simple cyclocondensation method and ring closing agents for synthesis of 2-substituted benzimidazoles.

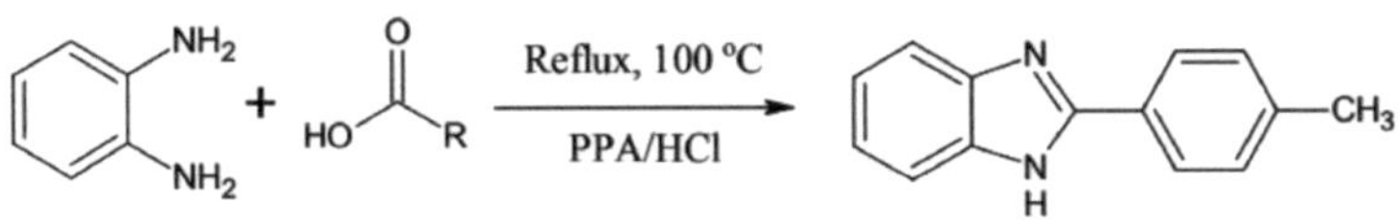

Rekha et al. [7] studied the catalytic activity of alumina, zirconia, manganese oxide/alumina and manganese oxide/zirconia in the condensation reaction between o-phenylenediamine and an aldehyde or a ketone to synthesize 2-substituted benzimidazoles and 1,5-disubstituted benzodiazepines, respectively.

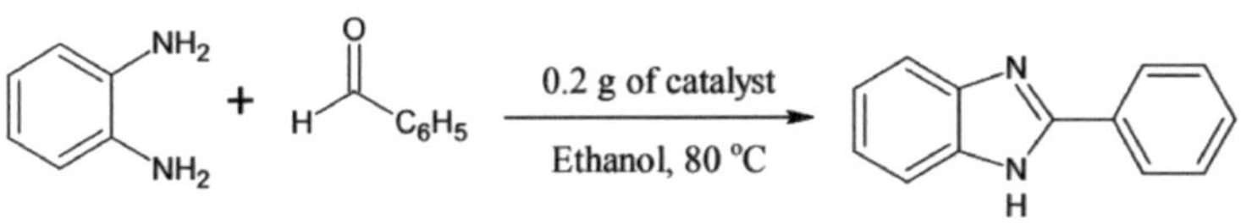

Chunxia et al. [8] have developed a straightforward method for the synthesis of the benzimidazole ring system through a carbon-nitrogen cross-coupling reaction in the presence of K_2CO_3 in water at 100°C for 30 h; the intermolecular cyclization of N-(2-iodoaryl) benzamidine provides benzimidazole derivatives in moderate to high yields.

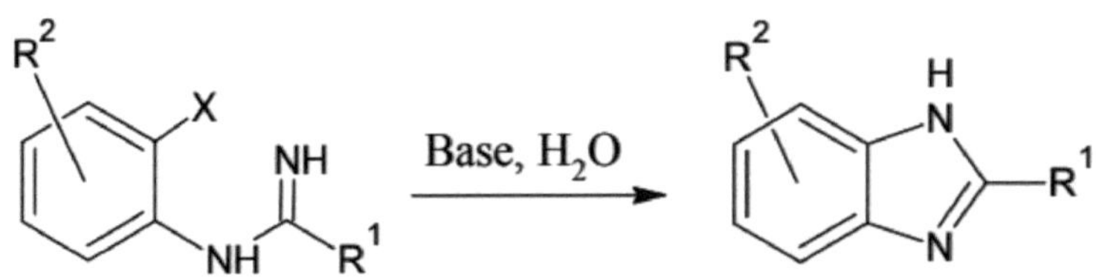

Kathirvelan et al. [9] synthesized various 2-substituted benzimidazoles in moderate to good yields in a one-pot reaction by condensation of o-phenylenediamine and an aldehyde in the presence of ammonium chloride as a catalyst at 80–90°C.

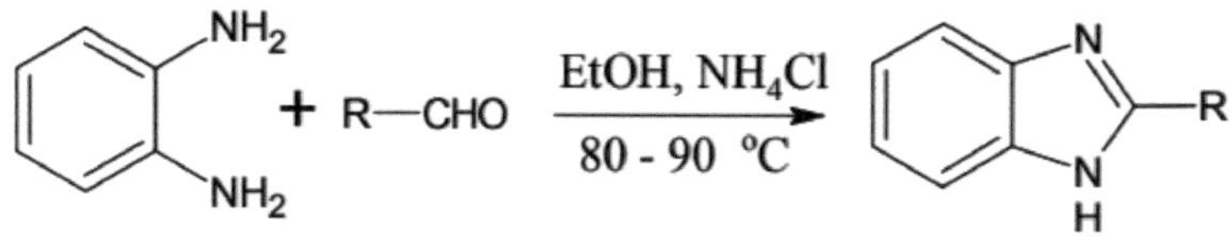

Srinivasulu et al. synthesized benzimidazole derivatives using zinc triflate as an efficient catalyst in a one-pot synthesis of 2-substituted benzimidazole derivatives from o-phenylenediamine and substituted aldehydes in ethanol solvent at reflux temperature. They concluded that zinc triflate was found to be an efficient catalyst for the formation of benzimidazole from aldehydes and o-phenylenediamine [10].

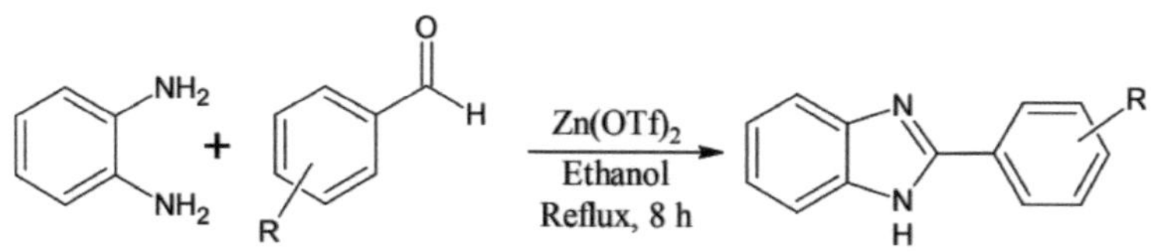

Patil et al. [11] have shown that benzimidazole derivatives have been synthesized by reacting substituted o-phenylenediamine with aldehyde derivatives using a catalytic amount of zinc acetate at room temperature with excellent yields.

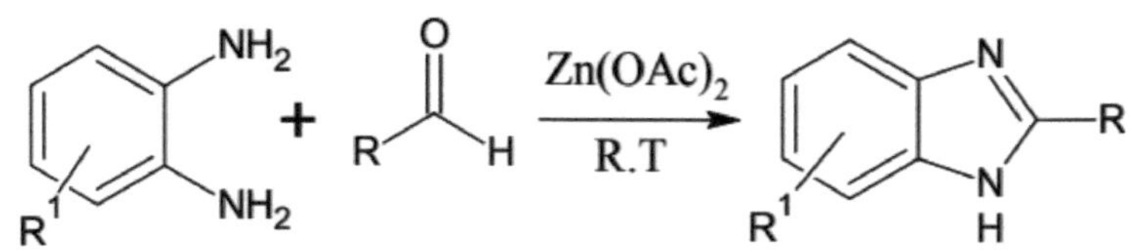

Karimi-Jaberi et al. [12] had synthesized 2-substituted benzimidazoles in a one-pot reaction from o-phenylenediamine and aldehydes in the presence of boric acid in water at room temperature.

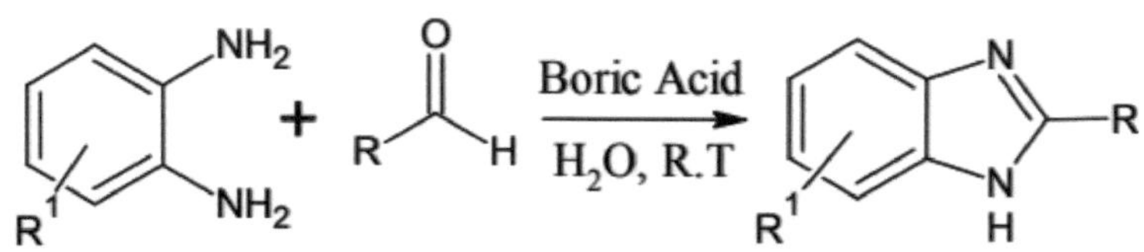

Kidwai et al. [13] synthesized benzimidazole derivatives from o-phenylenediamine and aldehydes in PEG as a solvent with ceric ammonium nitrate (CAN) as a catalyst. This method provides a novel route for the synthesis of benzimidazoles in good yields with little catalyst loading.

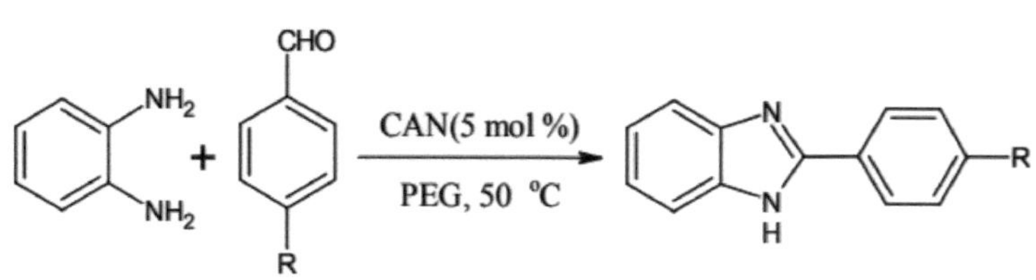

Khunt et al. [14] have synthesized the benzimidazole by reacting o-phenylenediamine with several aldehydes using a green solvent PEG400 and got good yields.

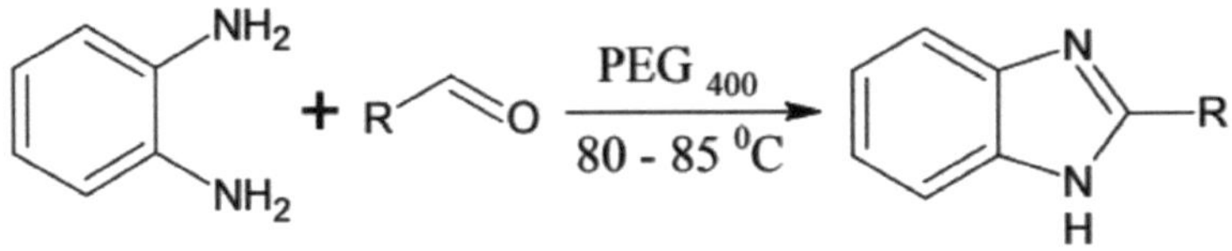

2. Corrosion inhibition by benzimidazole derivatives

The inhibition ability of the organic compounds containing heteroatoms depends upon the rate of the molecular adsorption at the metal/solution interface which in turn depends upon the molecular structure/symmetry and charge density presenting on the inhibitor molecules [15]. Benzimidazoles are very good corrosion inhibitors because of their unique molecular structure and solubility in aqueous media. As we have seen in **Figure 3**, the molecular structure of benzimidazole has an aromatic property due to the presence of the fused six-membered benzene ring with the imidazole ring. The presence of planar benzene ring with sp2 hybridized carbon atoms and lone pair of electrons presenting on the nitrogen atoms of the imidazole ring makes the whole molecular system as an anchoring site for molecular adsorption at the metal surface [16, 17].

2.1 Metal protection by benzimidazole in aqueous media

Benzimidazole and its derivatives prevail themselves as potential corrosion inhibitor candidates. They prove the ability to reduce the corrosion phenomenon in aqueous media for various metals like mild steel, copper, brass and zinc. In this chapter, I would like to emphasize only on these metals. The molecular structures of benzimidazole derivatives studied in this book chapter are tabulated in **Table 1**.

2.1.1 Experimental results

Ramya et al. [18] have studied the synergistic hydrogen-bonded interaction of alkyl benzimidazole derivatives (2-MBI, 2-EBI, 2-PBI), and 1,2,3-benzotrizole (BTZ) and its corrosion protection properties on mild steel in hydrochloric acid at different temperatures have been studied using polarization, EIS, adsorption, surface studies and computational methods. They observed that benzimidazole derivatives and BTZ molecules are effective inhibitors and their inhibition efficiency increases when these two inhibitors are used in combination.

In 2011 Guadalupe et al. [19] analysed the corrosion inhibition properties of BBED via electrochemical (polarization curves and electrochemical impedance spectroscopy) and DFT techniques. Electrochemical impedance data demonstrate that the interface charge capacitance of the electrode with the BBED solution affects the time of charge/discharge of the interface, facilitating the formation of adsorption layer over the iron surface. They also estimated the fractal dimension of the electrode surface in order to understand the nature of the electrode surface. DFT results clearly show that BBED possess corrosion inhibition properties by having a delocalization region (N1=C1=N2) in the benzimidazole ring that gives up its p

Structure	Metal/medium	Reference
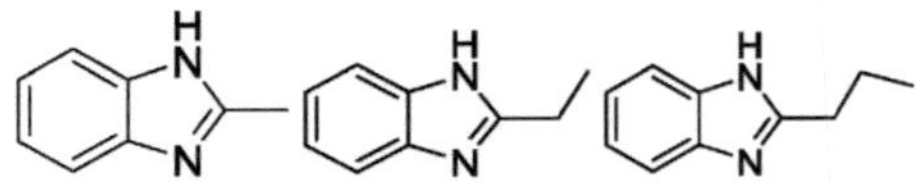 (2-MBI) (2-EBI) (2-PBI)	Mild steel/1 M HCl	[18]
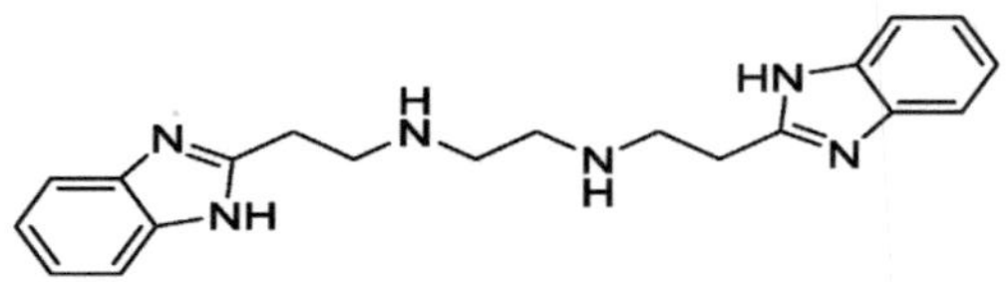 BBED	Carbon steel/1 M HCl	[19]
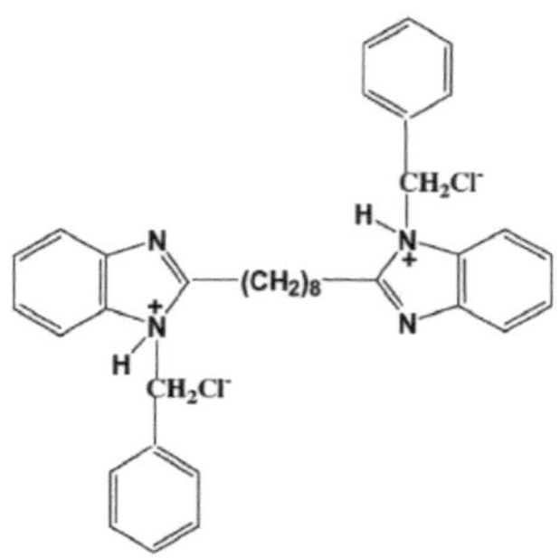 CBO	Mild steel/0.1–1 M HCl	[20]

Structure	Metal/medium	Reference
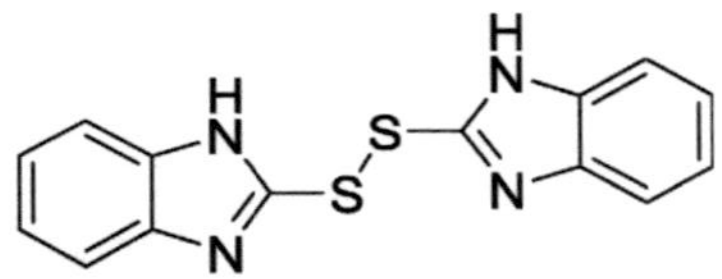 BIMDS	Mild steel/1 M HCl and 0.5 M H_2SO_4	[21]
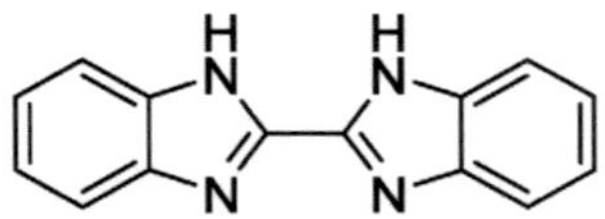 2,2'-bis(benzimidazole)	Mild steel/1 M HCl	[22]
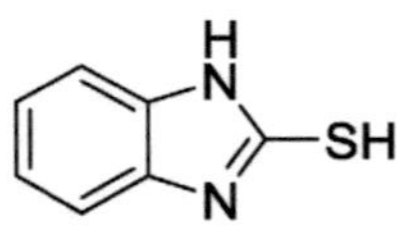 2-SH-BI 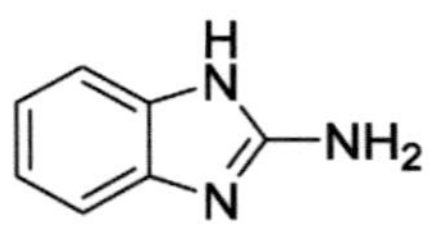 2-NH2-BI	Mild steel/1 M HCl	[23, 24]
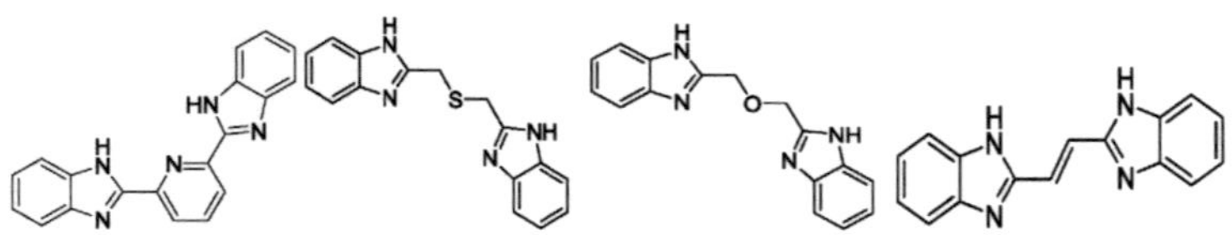 BBP BBE BBMS BBMO	Mild steel/1 M HCl	[25]

Structure	Metal/medium	Reference
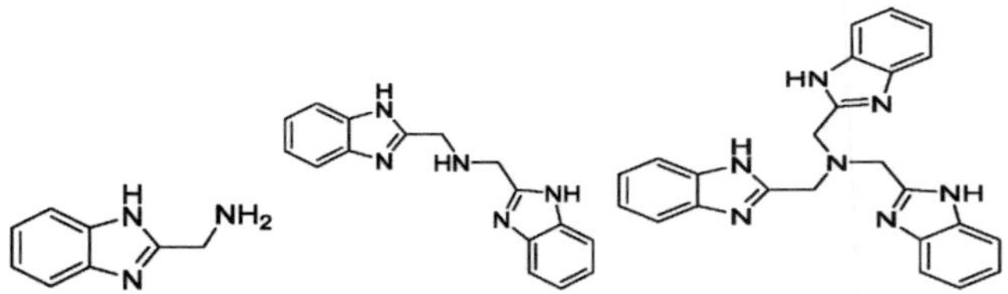 ABI BBIA TBIA	Mild steel/1 M HCl	[26]
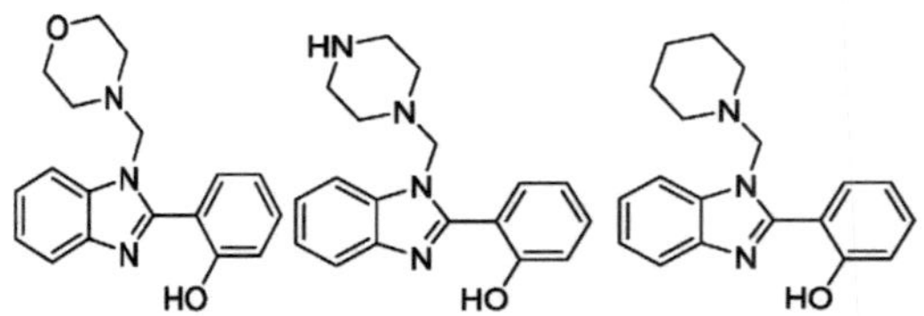 MBP PzMBP PMBP	N80 steel/15% HCl	[27]
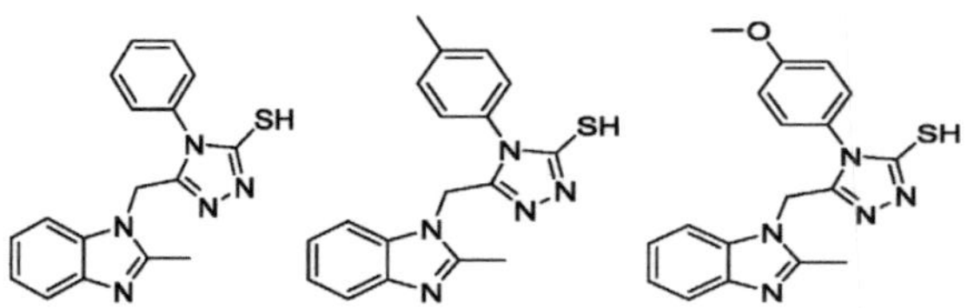 Inh I InhII InhIII	N80 steel/15% HCl	[28]
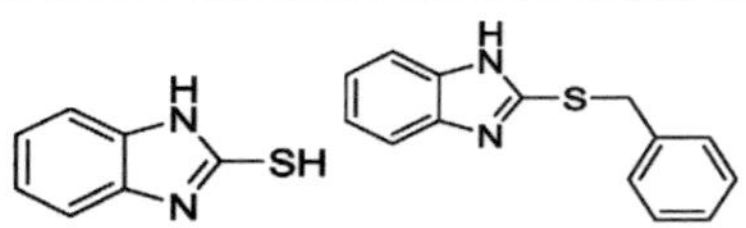 MBI TBBI	Cu/1 M HNO_3	[29]

Structure	Metal/medium	Reference
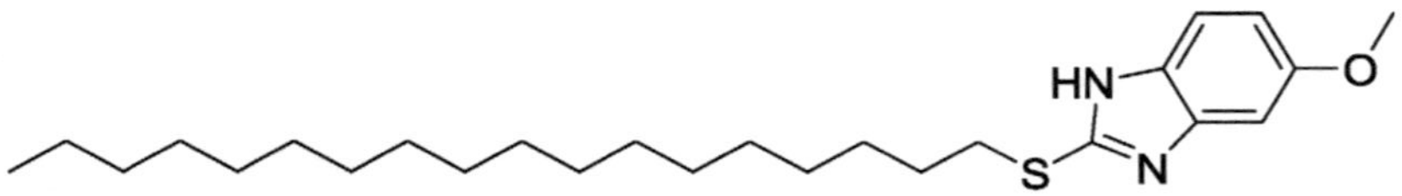 MOTBI	Cu/0.02 M NaCl	[30]
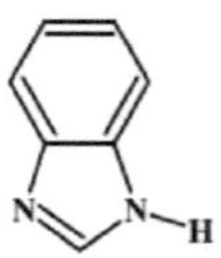 BI 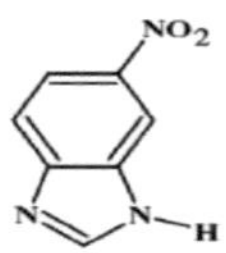 5-NO2BI 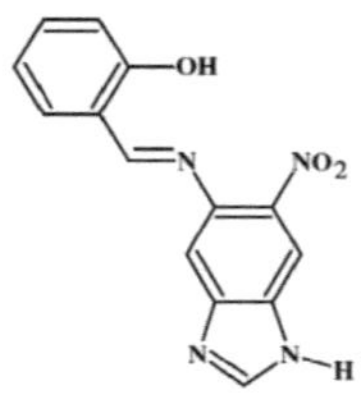 BSB 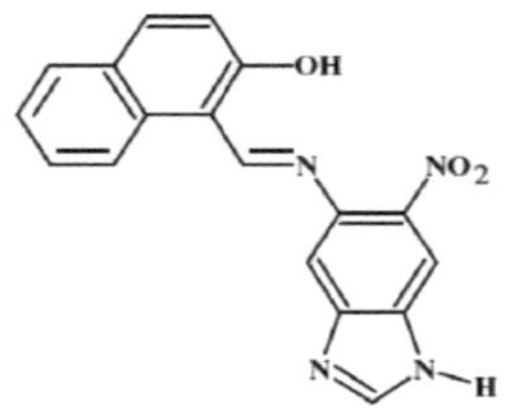 NSB	Cu and brass/0.4 M NaCl + 0.1 M NaOH	[31]
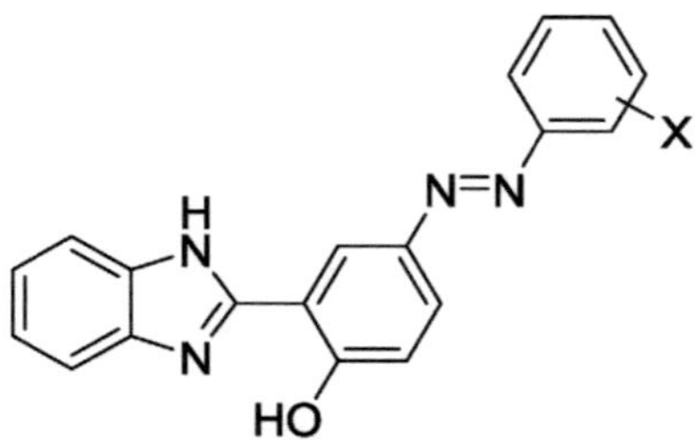 X=o-OCH_3, p-OCH_3, o-CH_3, m-CH_3, m-OCH_3, p-NO_2, m-NO_2 (BPP-1) (BPP-2) (BPP-3) (BPP-4) (BPP-5) (BPP-6) (BPP-7)	Cu/2 M HNO_3	[32]

Structure	Metal/medium	Reference
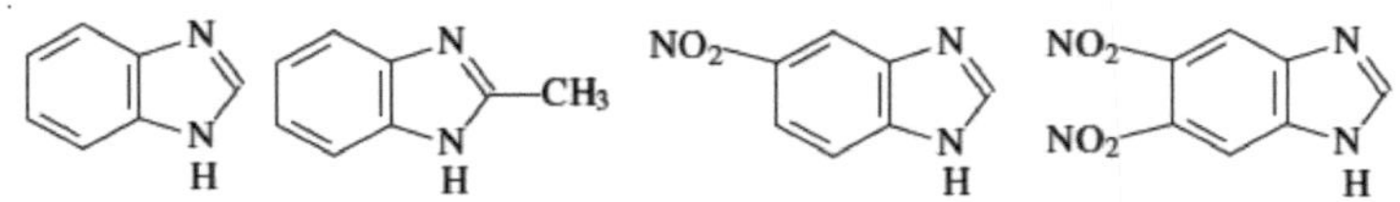 BIM 2-CH3BIM 5-NO2BIM 5,6-di NO2BIM	Copper, zinc, brass/0.5 M NaCl	[33]
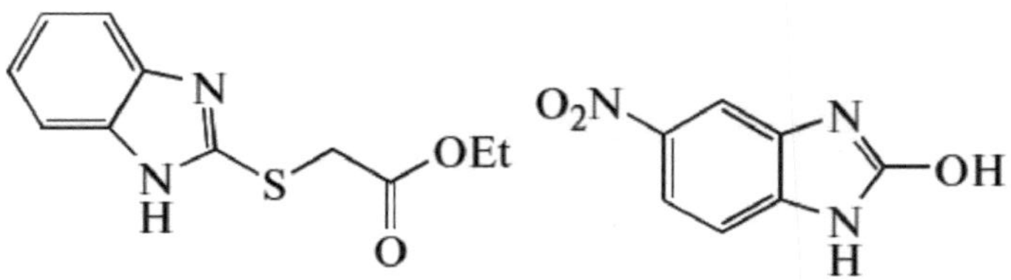 MBEA HBI	Zn/0.1 M HCl	[34]

Table 1.
The molecular structure of benzimidazole derivatives.

electron density through its HOMO orbital to the metal LUMO to form an adsorption layer over the metallic surface.

The influences of a benzimidazole derivative, namely, 1,8-bis(1-chlorobenzyl-benzimidazolyl)-octane (CBO), on the corrosion behaviour of mild steel in different concentration HCl solutions was studied by Wang et al. [20]. The authors conclude that CBO acted as an excellent mixed-type inhibitor and strongly adsorbed via chemical bond formation. The data obtained from weight loss and electrochemical measurements have shown that the CBO has the excellent inhibiting properties for mild steel in HCl solution.

Ahamad and Quraishi [21] studied the corrosion inhibition effects of a water-soluble benzimidazole derivative, namely, bis(benzimidazol-2-yl)disulphide (BIMDS) on mild steel 1.0 M HCl and 0.5 M H_2SO_4. The results reveal that BIMDS showed a better performance in 0.5 M H_2SO_4 solutions than in 1.0 M HCl. Polarization measurements indicated that BIMDS is a mixed-type inhibitor in both acid media. The inhibitor obeyed the Langmuir adsorption isotherm model in both acid media.

Abboud et al. [22] studied the 2,2'-bis(benzimidazole) as a corrosion inhibitor for mild steel in 1 M HCl. The analysis of polarization curves showed that the mechanism of corrosion inhibition remains unchanged after the addition of the inhibitor and the inhibitor acts as a mixed type. The corrosion inhibition of the inhibitor can be interpreted by a simple blocked fraction of the electrode surface related to the adsorption of the inhibitor species according to a Langmuir isotherm on the steel surface.

Popova et al. and Mahdavian et al., respectively, in 2004 and 2010 studied the corrosion inhibition property of a series of benzimidazole derivatives [23, 24]. The results of Popova et al. reveal that the inhibition efficiency increased with the increase of organic substrate concentration, while the adsorption followed the Frumkin isotherm. They also suggest that there was no correlation observed among the studied various parameters of the electronic structure and the corrosion inhibition efficiency. Likewise Mahdavian results reveal that the studied inhibitor reduced both the cathodic and anodic polarization curves and behaved as a mixed-type inhibitor. The change in the values of the impedance parameters suggests the adsorption of the inhibitor on the metal surface. In surface analysis like SEM/EDX, the presence of sulphur confirmed the adsorption of the inhibitor on the mild steel surface.

Dutta et al. [25] in 2015 studied the corrosion inhibition of bis-benzimidazole type derivatives for mild steel with an immersion of 4 days in 1 M HCl. The results of the potentiodynamic polarization studies reveal that inhibitors produce a mixed-type protection but mostly affect the cathodic reaction.

2.1.2 Computational results

The approach of DFT results were used to probe into the interaction modes for the inhibitor molecules on the steel surface. The MDS was used to analyse the configuration of equilibrium adsorption of inhibitors. All the interaction energies between the inhibitors and the metal surface were deduced in the aqueous phase.

Cao et al. [26] performed the comparative theoretical study using DFT and MSD for 2-aminomethyl benzimidazole (ABI), bis(2-benzimidazolylmethyl) amine (BBIA) and tri-(2-benzimidazolylmethyl) amine (TBIA) on the mild steel surface. They concluded that the three-inhibitor molecules showed similar ability for the electron donation, while the difference in the electron-accepting ability makes them give a different inhibition performance. MD simulations suggest that the steric effect among the inhibitor molecular structures significantly affects the adsorptive

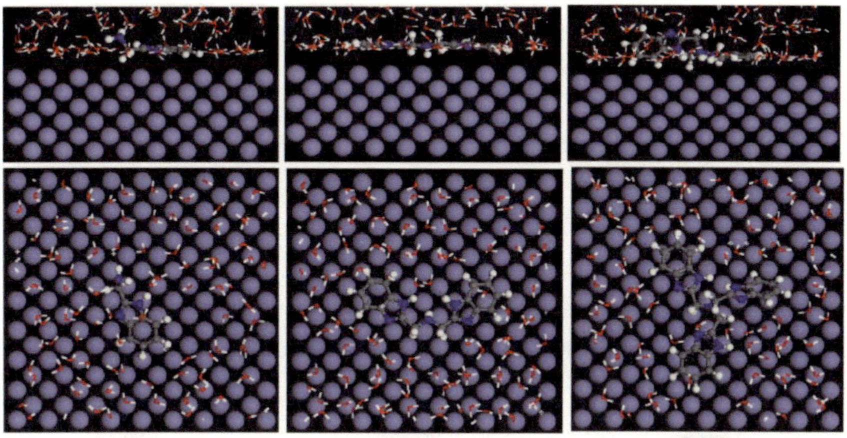

Figure 4.
Equilibrium configurations of ABI, BBIA and TBIA molecules.

configurations over the Fe (1 0 0) surface. **Figure 4** represents the equilibrium adsorption configurations of inhibitors on Fe (1 0 0) in the aqueous phase.

Yadav et al. [27, 28] group studied the corrosion inhibition of 2-(1-(morpholinomethyl)-1Hbenzo[d]imidazol-2-yl)phenol (MBP), 2-(1-((piperazine-1-yl) methyl)-1H-benzo[d]imidazol-2-yl)phenol (PzMBP), 2-(1-((piperidine-1-yl) methyl)-1H-benzo[d]imidazol-2-yl)phenol (PMBP), 4-(phenyl)-5-[(2-methyl-1H-benzimidazol-1-yl)methyl]-4H-1,2,4-triazole-3-thiol (Inh I), 4-(4-methylphenyl)-5-[(2-methyl-1H-benzimidazol-1-yl)methyl]-4H-1,2,4- triazole-3-thiol (Inh II) and 4-(4-methoxyphenyl)-5-[(2-methyl-1H-benzimidazol-1-yl)methyl]-4H-1,2,4-triazole-3-thiol (Inh III) on N80 steel in 15% HCl solution using weight loss measurement, potentiodynamic polarization and electrochemical impedance spectroscopy (EIS) techniques. They observed that inhibition efficiency of all the inhibitors increases with increase in concentration and decreases with increase in temperature. Furthermore the steel surface was analysed by SEM and AFM techniques. The density functional theory was employed for the theoretical evaluation of the studied inhibitors. **Figures 5** and **6** represent the AFM and SEM micrograph of the N80 steel surface.

The inhibitive actions of 2-mercaptobenzimidazole (MBI) and 2 thiobenzylbenzimidazole (TBBI) on copper corrosion in 1 M HNO_3 medium were studied, using weight loss method, at 25–65°C and in the concentrations of the range of 5×10^{-5} M to 10^{-3} M by Niamien et al. group in 2011 [29]. They reported the inhibition efficiencies of 90.0% for TBBI and 87.7% for MBI at 25°C. Negative values of changes in free energies indicate the spontaneous adsorption process with a combination of both physical and chemical processes. They have done the correlation between theoretical data and experimental results using DFT/B3LYP/6-31G (d,p) method.

Rao et al. [30] demonstrated the self-assembled (SAM) monolayer formation of 5-methoxy-2-(octadecylthio)benzimidazole (MOTBI) on fresh copper surface. The MOTBI SAM on copper surface was characterized by contact angle measurements, X-ray photoelectron spectroscopy and reflection absorption FTIR spectroscopy, and it is inferred that chemisorption of MOTBI on copper surface is through nitrogen. The corrosion protection ability of MOTBI SAM was evaluated in aqueous NaCl solution using impedance, electrochemical quartz crystal nanobalance, potentiodynamic polarization and weight loss studies.

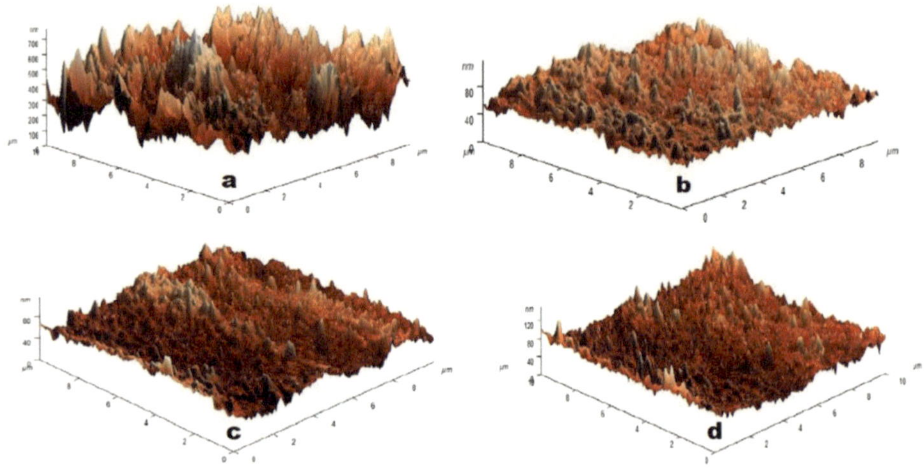

Figure 5.
AFM micrograph (a) without inhibitor, (b) with PzMBP, (c) with MBP and (d) with PMBP.

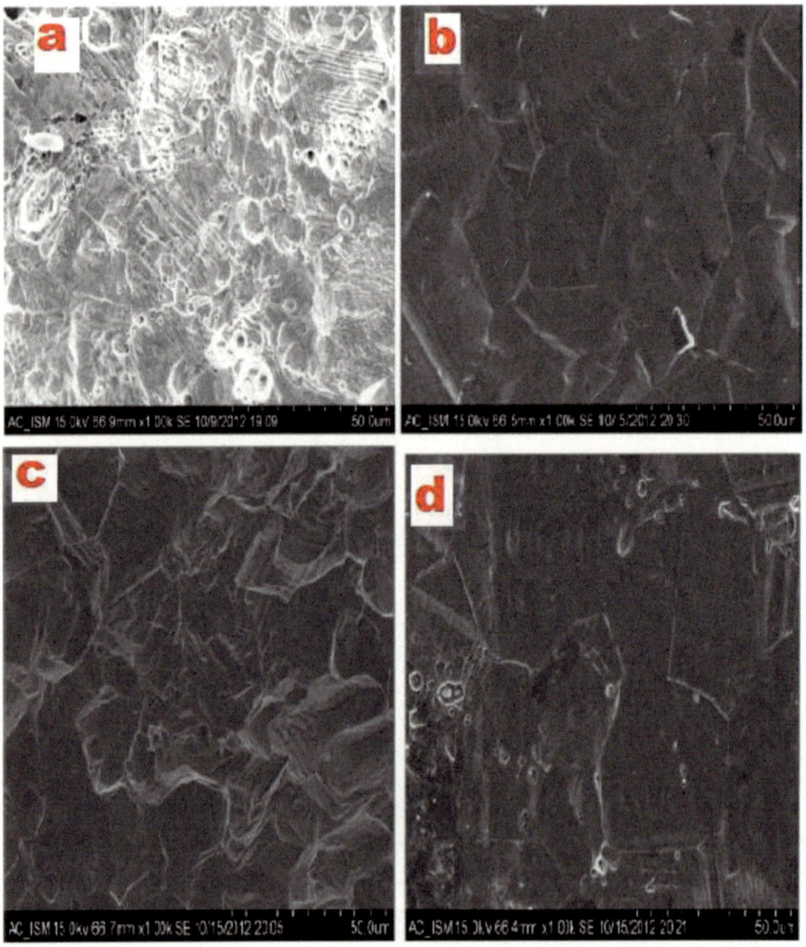

Figure 6.
SEM micrograph (a) without inhibitor (b) Inh III, (c) Inh II and (d) Inh I.

In 2014, Özbay et al. [31] have carried out the inhibitive action of newly synthesized ortho-hydroxy Schiff bases of 5-amino-6-nitro-1H-benzimidazole against the corrosion of copper and brass in alkaline medium using potentiodynamic polarization and electrochemical impedance spectroscopy. The results showed an inhibition efficiency of 91.0% and 97.4% for copper and brass, respectively.

Madkour and Elsham [32] studied the inhibitive performance of seven synthesized 2-(2-benzimidazolyl)-4 (phenylazo) phenol (BPP_1–7) derivatives which were investigated experimentally on the corrosion of copper in 2.0 M HNO_3 acid using mass loss and thermometric and DC potentiodynamic polarization techniques. Quantum chemical calculations were investigated to correlate the electronic structure parameters of the investigated benzimidazole derivatives with their inhibition efficiencies values.

Yanardag et al. [33] have studied the corrosion inhibition property of 1H-benzimidazole, (BIM), 2-methyl-1H-benzimidazole (2-CH_3BIM), 5-nitro-1H-benzimidazole [5(6)-NO_2BIM] and 5(6)-dinitrobenzimidazole (5,6-diNO_2BIM) for copper, zinc and brass in alkaline and neutral media. The results under the experimental conditions suggest the efficiency order of the inhibitors is BIM > 5(6)-NO_2BIM > 5,6-diNO_2BIM > 2-CH_3BIM in alkaline media (pH: 13) and 2-CH_3BIM > 5(6)-NO_2BIM > BIM in the 0.5 M NaCl solution.

The corrosion inhibition characteristics of 2-mercaptobenzimidazole (MBI), 2-mercapto benzimidazolyl-ethyl acetate (MBEA), 2-hydroxy benzimidazole (HBI) and 2-hydroxy 5-nitro benzimidazole (HNBI) on zinc corrosion in 0.1 M hydrochloric acid were investigated by weight loss and potentiodynamic polarization techniques by the group of Shanbhag et al. [34]. The inhibition efficiency

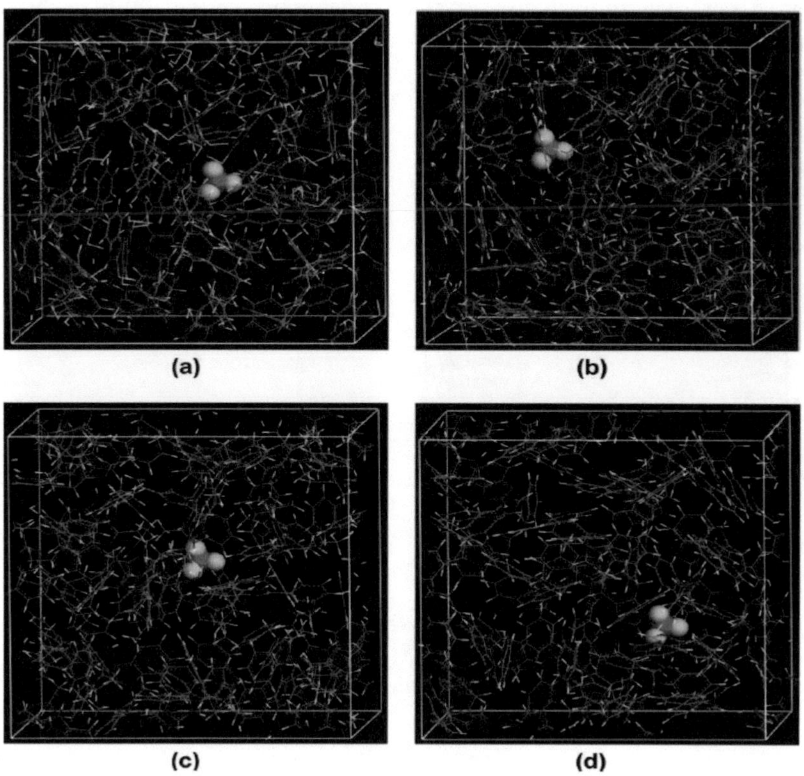

Figure 7.
Simulation models of H_3O^+ (shown in CPK) in various inhibitor films (shown in line): (a) 2-SH-BI, (b) 2-NH_2-BI, (c) 2-CH_3-BI and (d) BI.

increased with an increase in inhibitor concentration but decreased with an increase in temperature. The adsorption of MBI and MBEA obeyed Langmuir's adsorption isotherm, but HBI and HNBI followed Temkin's adsorption isotherm. The existence of inhibitor film on metal surface was established by scanning electron microscopy (SEM) images.

Zhang et al. [35] studied the diffusion of corrosive particles inside inhibitor films consisting of 2-mercaptobenzimidazole (2-SH-BI), 2-aminobenzimidazole (2-NH_2-BI), 2-methylbenzimidazole (2-CH_3-BI) and benzimidazole using molecular dynamics simulation (MDS). Diffusion coefficients of various corrosive particles in the films were calculated, and their order is 2-SH-BI < 2-NH_2-BI < 2-CH_3-BI < BI. Fractional free volume, interaction between corrosive particles and films and mobility of films were also investigated to illustrate the microscopic diffusion mechanism. As a result, it can be inferred that the order of inhibition efficiency is 2-SH-BI > 2-NH_2-BI > 2-CH_3-BI > BI. The simulation models of H_3O^+ in various inhibitor films are shown in **Figure 7**.

Recently Singh et al. [36] have studied the corrosion inhibition effect of three synthesized benzimidazole derivatives, namely, 2-(3,4,5-trimethoxyphenyl)-1H-benzo[d] imidazole (TMI), 2-(3,4-dimethoxyphenyl)-1H-benzo[d] imidazole (DMI) and 2-(4-methoxyphenyl)-1H-benzo[d] imidazole (MMI), for J55 steel saturated with CO_2 in a 3.5% NaCl solution. The author has analysed the corrosion inhibition property using weight loss, impedance spectroscopy (EIS) and potentiodynamic polarization methods. The metal surface was examined by scanning electron microscope (SEM) and X-ray photoelectron spectroscopy (XPS). They have justified the experimental results using the DFT and MD studies. Their investigations both by experimental and theoretical analyses suggest that as the number of methoxy groups increases, the corrosion protection ability of the inhibitors increases, and thus TMI is the best inhibitor. The results of the Tafel curves are presented in **Figure 8**. As observed from the curves, inhibitor TMI exhibited a

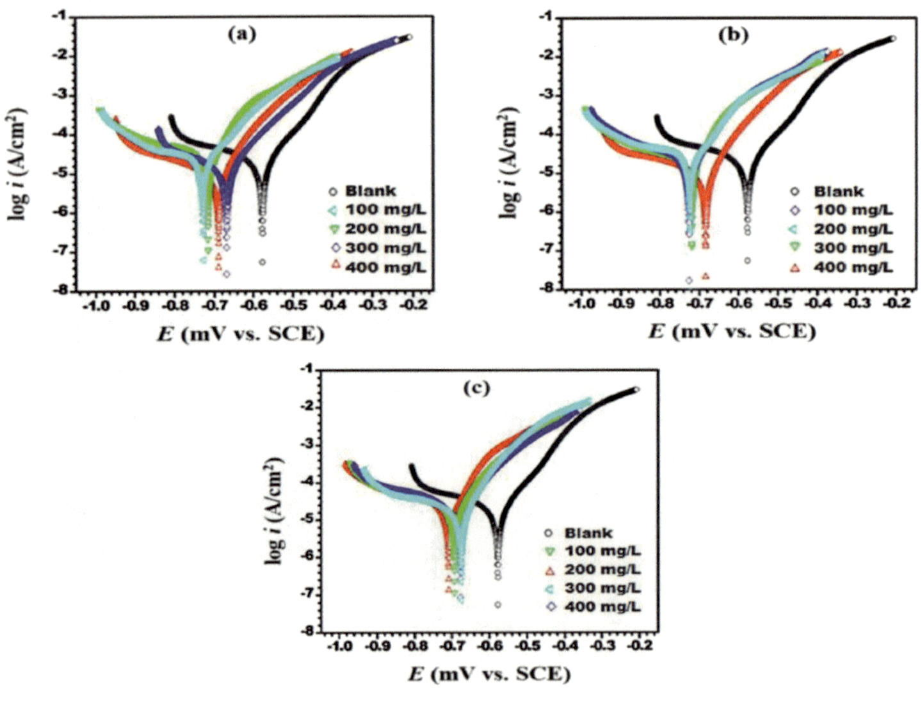

Figure 8.
Tafel curves in without and with different concentrations of inhibitors: (a) TMI; (b) DMI; (c) MMI.

superior inhibition as compared to DMI and MMI, and the adsorption of these molecules on the metal surface affected both anodic and cathodic reactions. The three studied compounds can significantly reduce the corrosion current density values, and they exhibit mixed-type inhibition characteristics. The values of corrosion current density (icorr) without addition inhibitors are 104.4 μA cm^{-2}, and the addition of inhibitors reduced it to 5.5 μA cm^{-2} (TMI), 13.2 μA cm^{-2} (DMI) and 21.5 μA cm^{-2} (MMI).

3. Conclusion

The book chapter gives an overview of the role of benzimidazole and their derivatives in the corrosion mitigation of industrially important alloys such as mild steel, N80 steel, aluminium, zinc, copper and brass in various aggressive media. Here we have also tried to explore the molecular structure, synthesis protocol and some important biological activity of benzimidazole derivatives. In view of the literature analysis, it could be concluded that benzimidazole derivatives have the potential ability to act as a suitable additive for various corrosive solutions.

Acknowledgements

The author Dr. Ambrish Singh is thankful to the Sichuan 1000 Talent Fund and the National Natural Science Foundation of China (No. 51274170) for providing financial assistance.

Author details

Ambrish Singh[1*], Kashif R. Ansari[2], Dheeraj S. Chauhan[2], Mumtaz A. Quraishi[2], Savas Kaya[3], Hua Yu[4] and Yuanhua Lin[1]

1 School of New Energy and Materials, Southwest Petroleum University, Chengdu, Sichuan, China

2 Centre of Research Excellence in Corrosion, Research Institute, King Fahd University of Petroleum and Minerals, Dhahran, Saudi Arabia

3 Department of Chemistry, Sivas Cumhuriyet University, Turkey

4 Institute of Photovoltaics, Southwest Petroleum University, Chengdu, Sichuan, China

*Address all correspondence to: vishisingh4uall@gmail.com

IntechOpen

© 2020 The Author(s). Licensee IntechOpen. This chapter is distributed under the terms of the Creative Commons Attribution License (http://creativecommons.org/licenses/by/3.0), which permits unrestricted use, distribution, and reproduction in any medium, provided the original work is properly cited.

References

[1] Negi DS, Kumar G, Singh M, Singh N. Antibacterial activity of benzimidazole derivatives: A mini review. Research & Reviews: Journal of Chemistry. 2017;**6**:18-28

[2] Curkovic HO, Stupnisek-Lisac E, Takenouti H. The influence of pH value on the efficiency of imidazole based corrosion inhibitors of copper. Corrosion Science. 2010;**52**: 398-405

[3] Infante-Castillo R, Hernández-Rivera SP. Experimental and theoretical studies of the molecular structure of 1-(2-pyridinylmethyl)-2-methylbenzimidazole. Journal of Molecular Structure. 2010;**968** (1-3):6-12

[4] Azarifar D, Pirhayati M, Maleki B, Sanginabadi M, Yami RN. Acetic acid-promoted condensation of o-phenylenediamine with aldehydes into 2-aryl-1 (arylmethyl)-1H-benzimidazoles under microwave irradiation. Journal of Serbian Chemical Society. 2010;**75**:1181-1189

[5] Shaikh KA. An efficient solvent-free synthesis of imidazolines and benzimidazoles using $K_4[Fe(CN)_6]$ catalysis. Organic Communication. 2012;**5**:12-17

[6] Vaidehi BNB. Synthesis and biological evaluation of 2-substituted benzimidazoles. International Journal of Pharma and Biosciences. 2012;**3**:26-31

[7] Rekha M, Hamza A, Venugopal BR, Nagaraju N. Synthesis of 2-substituted benzimidazoles and 1,5-disubstituted benzodiazepines on alumina and zirconia catalysts. Chinese Journal of Catalysis. 2012;**33**:439-446

[8] Chunxia C, Chen C, Bin L, Jingwei T, Jinsong P. Aqueous synthesis of 1-H-2-substituted benzimidazoles via transition-metal-free intramolecular amination of aryl iodides. Molecules. 2012;**1**:12506-12520

[9] Kathirvelan D, Yuvaraj P, Babu K, Nagarajan AS, Reddy BSR. A green synthesis of benzimidazoles. Indian Journal of Chemistry. 2013;**52B**: 1152-1156

[10] Srinivasulu R, Kumar KR, Satyanarayana PVV. Facile and efficient method for synthesis of benzimidazole derivatives catalyzed by zinc triflate. Green and Sustainable Chemistry. 2014; **4**:33-37

[11] Patil VD, Medha G, Shramesha M, Aarti J. A mild and efficient synthesis of benzimidazole by using lead peroxide under solvent free condition. Der Chemica Sinica. 2010;**1**:125-129

[12] Karimi-Jaberi Z, Amiri M. An efficient and inexpensive synthesis of 2-substituted benzimidazoles in water using boric acid at room temperature. E-Journal of Chemistry. 2012;**9**: 167-170

[13] Kidwai M, Jahan N, Bhatnagar D. Polyethylene glycol: A recyclable solvent system for the synthesis of benzimidazole derivatives using CAN as catalyst. Journal of Chemical Sciences. 2010;**122**:607-612

[14] Khunt MD, Kotadiya VC, Viradiya DJ, Baria BH, Bhoya UC. Easy, simplistic and green synthesis of various benzimidazole and benzoxazole derivatives using PEG400 as a green solvent. International Letters of Chemistry, Physics and Astronomy. 2014;**25**:61-68

[15] Aljourani J, Raeissi K, Golozar MA. Benzimidazole and its derivatives as corrosion inhibitors for mild steel in 1 M HCl solution. Corrosion Science. 2009; **51**:1836-1843

[16] Cruz J, Martínez R, Genesca J, García-Ochoa E. Experimental and theoretical study of 1-(2-ethylamino)-2-methylimidazoline as an inhibitor of carbon steel corrosion in acid media. Journal of Electroanalytical Chemistry. 2004;**566**:111-121

[17] Khaled KF. The inhibition of benzimidazole derivatives on corrosion of iron in 1 M HCl solutions. Electrochimica Acta. 2003;**48**:2493-2503

[18] Ramya K, Mohan R, Joseph A. Interaction of benzimidazoles and benzotriazole: Its corrosion protection properties on mild steel in hydrochloric acid. Journal of Materials Engineering and Performance. 2014;**23**:4089-4101

[19] Guadalupe HJ, García-Ochoa E, Maldonado-Rivas PJ, Cruz J, Pandiyan T. A combined electrochemical and theoretical study of N,N'-bis(benzimidazole-2yl-ethyl)-1,2-diaminoethane as a new corrosion inhibitor for carbon steel surface. Journal of Electroanalytical Chemistry. 2011;**655**:164-172

[20] Wang X, Yang H, Wang F. An investigation of benzimidazole derivative as corrosion inhibitor for mild steel in different concentration HCl solutions. Corrosion Science. 2011; **53**:113-121

[21] Ahamad I, Quraishi MA. Bis (benzimidazol-2-yl) disulphide: An efficient water soluble inhibitor for corrosion of mild steel in acid media. Corrosion Science. 2009;**51**:2006-2013

[22] Abboud Y, Abourriche A, Saffaj T, Berrada M, Charrouf M, Bennamara A, et al. The inhibition of mild steel corrosion in acidic medium by 2,2'-bis (benzimidazole). Applied Surface Science. 2006;**252**:8178-8184

[23] Popova A, Christov M, Raicheva S, Sokolova E. Adsorption and inhibitive properties of benzimidazole derivatives in acid mild steel corrosion. Corrosion Science. 2004;**46**:1333-1350

[24] Mahdavian M, Ashhari S. Corrosion inhibition performance of 2-mercaptobenzimidazole and 2-mercaptobenzoxazole compounds for protection of mild steel in hydrochloric acid solution. Electrochimica Acta. 2010;**55**:1720-1724

[25] Dutta A, Saha SK, Banerjee P, Sukul D. Correlating electronic structure with corrosion inhibition potentiality of some bis-benzimidazole derivatives for mild steel in hydrochloric acid: Combined experimental and theoretical studies. Corrosion Science. 2015;**98**:541-550

[26] Cao Z, Tang Y, Cang H, Xu J, Lu G, Jing W. Novel benzimidazole derivatives as corrosion inhibitors of mild steel in the acidic media. Part II: Theoretical studies. Corrosion Science. 2014;**83**:292-298

[27] Yadav M, Kumar S, Purkait T, Olasunkanmi LO, Bahadur I, Ebenso EE. Electrochemical, thermodynamic and quantum chemical studies of synthesized benzimidazole derivatives as corrosion inhibitors for N80 steel in hydrochloric acid. Journal of Molecular Liquids. 2016;**213**:122-138

[28] Yadav M, Behera D, Kumar S, Sinha RR. Experimental and quantum chemical studies on the corrosion inhibition performance of benzimidazole derivatives for mild steel in HCl. Industrial & Engineering Chemistry Research. 2013;**52**:6318-6328

[29] Niamien PM, Essy FK, Trokourey A, Sissouma D, Diabate D. Inhibitive effects of 2-mercaptobenzimidazole (MBI) and 2-thiobenzylbenzimidazole (TBBI) on copper corrosion in 1 M nitric acid solution. African Journal of Environmental Science and Technology. 2011;**5**:641-652

[30] Rao BVA, Iqbal MY, Sreedhar B. Electrochemical and surface analytical studies of the self-assembled monolayer of 5-methoxy-2-(octadecylthio) benzimidazole in corrosion protection of copper. Electrochimica Acta. 2011;**55**: 620-631

[31] Özbay S, Yanardağ T, Dinçer S, Aksüt AA. Benzimidazole Schiff bases as corrosion inhibitors for copper and brass. European International Journal of Science and Technology. 2014;**3**:1-6

[32] Madkour LH, Elshamy IH. Experimental and computational studies on the inhibition performances of benzimidazole and its derivatives for the corrosion of copper in nitric acid. International Journal of Industrial Chemistry. 2016;**7**:195-221

[33] Yanardag T, Özbay S, Dincer S, Aksut AA. Corrosion inhibition efficiency of benzimidazole and benzimidazole derivatives for zinc, copper and brass. Asian Journal of Chemistry. 2012;**24**:47-52

[34] Shanbhag AV, Venkatesha TV, Praveen BM. Benzimidazole derivatives as corrosion inhibitors for zinc in acid solution. Protection of Metals and Physical Chemistry of Surfaces. 2013; **49**:587-590

[35] Zhang J, Yu W, Yu L, Yan Y, Qiao G, Hu S, et al. Molecular dynamics simulation of corrosive particle diffusion in benzimidazole inhibitor films. Corrosion Science. 2011;**53**: 1331-1336

[36] Singh A, Ansari KR, Quraishi MA, Lgaz H. Effect of electron donating functional groups on corrosion inhibition of J55 steel in a sweet corrosive environment: Experimental, density functional theory, and molecular dynamic simulation. Materials 2019:**12**(1):17. Available from: https://doi.org/10.3390/ma12010017

Chapter 2

Green Inhibitors to Reduce the Corrosion Damage

Said Abbout

Abstract

Over the last years, corrosion phenomenon is an important economical and lives lost, which calls in the last decades researches on its final resolution by various techniques. Through this book chapter, we present one of the most used methods to protect the metals: the corrosion inhibitor. We have presented the classification (liquid and gas phase), action mode (adsorption, barrier, reinforcing of the oxide layer, passivation, and formed insoluble complex), the application fields (water treatment, petroleum industry), and some particular inhibitors. In addition, we present a case study using a green corrosion inhibitor (GCI) prepared from the oil of *Ceratonia siliqua* L. seeds.

Keywords: corrosion inhibitor, volatile corrosion inhibitors, *Ceratonia siliqua* L., polarization and impedance measurements

1. Introduction

Corrosion is an unstoppable phenomenon, in order to avoid or reduce the corrosion of metallic materials; the corrosion inhibitor is one of the most effective and flexible means of corrosion prevention and mitigation [1].

2. Generalities about the corrosion inhibitors

2.1 Definition and functions necessary in the corrosion inhibitor

According to ISO 8044, the corrosion inhibitor is a chemical substance added to the corrosion system at a concentration chosen for its effectiveness; this causes a decrease in the corrosion rate of the metal without significantly modifying the concentration of any corrosive agent contained in the aggressive medium [2]. In addition, this role can be assured by other ways such as modification of the pH and incorporation of some metals like zinc in the chemical composition of the materials. In fact, such a definition cannot be perfect; however, it avoids to consider inhibitors as additives.

From this definition, a corrosion inhibitor must therefore verify some fundamental properties:

- Decreasing the corrosion rate of the metal while retaining the physicochemical characteristics of the latter

- Stable in the presence of other constituents, in particular with respect to certain biocides
- Stable in the temperature range used
- Effective at low concentrations
- Inexpensive compared to the savings it allows to achieve
- Compatible with the current standards of nontoxicity and environmental pro-tection

2.2 Utilization conditions

A corrosion inhibitor can be used as method of protection:

- As a *permanent protection*, the inhibitor allows the use of metallic materials (non-alloyed ferrous example) under satisfactory conditions of resistance to corrosion [3].
- As a *temporary protection* during a period when the piece or installation is particularly susceptible to corrosion (storage, stripping, cleaning) [4]. In this case, the control of the system is a priori simpler, and the prediction of the behavior of the inhibitor over the time being is easier to carry out.
- As a *supplementary protection* to improve the resistance against the corrosion, in the case of the surface coating [5].

3. Classification of the corrosion inhibitors

Various authors have classified the corrosion inhibitors differently. Some authors prefer to group the inhibitors by their chemical functionality (organic or inorganic), others by their electrochemical reaction. **Figure 1** presents the classification of the corrosion inhibitors.

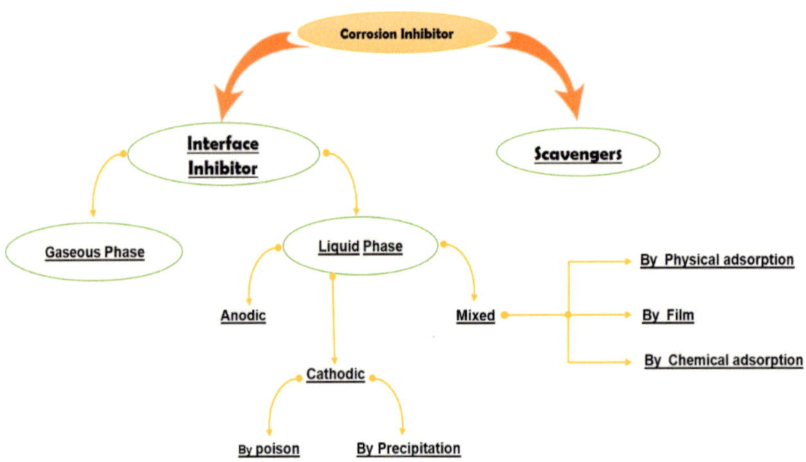

Figure 1.
Corrosion inhibitor classification.

3.1 Liquid phase

The corrosion inhibitors can be classified in the liquid phase as a cathodic, anodic, or mixed.

Cathodic inhibitors either slow the cathodic reaction itself or selectively precipitate on cathodic areas to increase the cathodic surface and limit the diffusion of aggressive species to these areas, which means a shift of the corrosion potential to a more negative potential (more than 85 mV) and a decrease in the cathodic current density. Cathodic inhibitors can provide inhibition by two different mechanisms: as cathodic poisons and as cathodic precipitates. Some cathodic inhibitors, such as arsenic and antimony compounds, work by making the recombination and discharge of hydrogen more difficult. Other cathodic inhibitors, such as calcium ions, zinc ions, or magnesium ions, may be precipitated as oxides to form a protective layer on the metal. Oxygen scavengers help to inhibit corrosion by preventing the cathodic depolarization caused by oxygen. The most commonly used oxygen scavenger at ambient temperature is probably sodium sulfite (Na_2SO_3) [6] (**Figure 2**).

Anodic (or passivating) *inhibitors* cause a large anodic shift of the corrosion potential to a more positive potential (more than 85 mV), forcing the metallic surface into the passivation range (slow the anodic reaction). There are two types of passivating inhibitors: oxidizing anions, such as chromate and nitrate, which can passivate steel in the absence of oxygen and the nonoxidizing ions, such as phosphate, tungstate, and molybdate, which require the presence of oxygen to passivate steel [7].

Anodic inhibitors can actually cause pitting and accelerate corrosion when concentrations fall below minimum limits. This kind of problem is not encountered in the case of cathodic inhibitors (**Figure 3**).

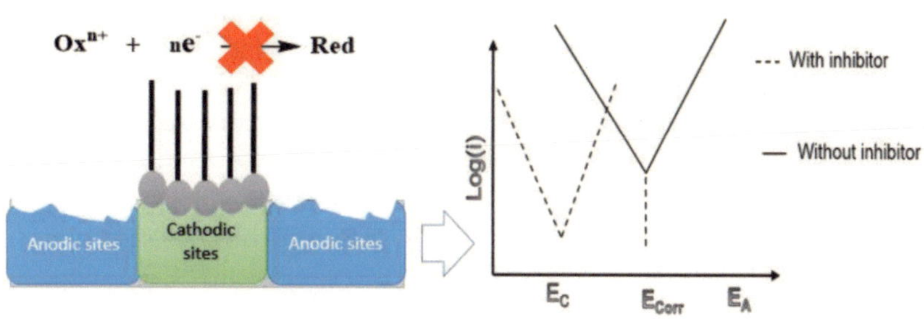

Figure 2.
Effect of addition of the cathodic inhibitor.

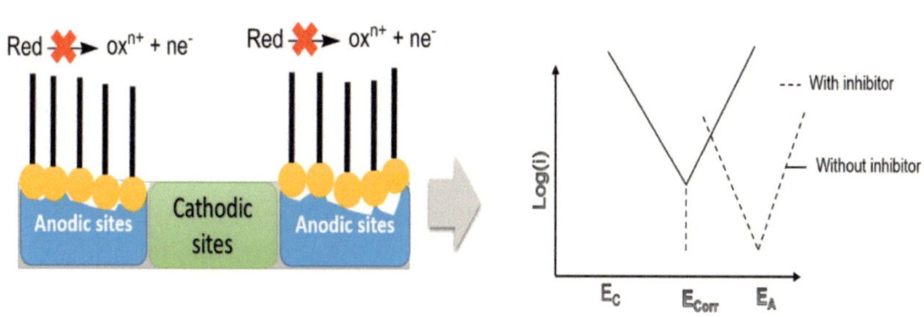

Figure 3.
Effect of addition of the anodic inhibitor.

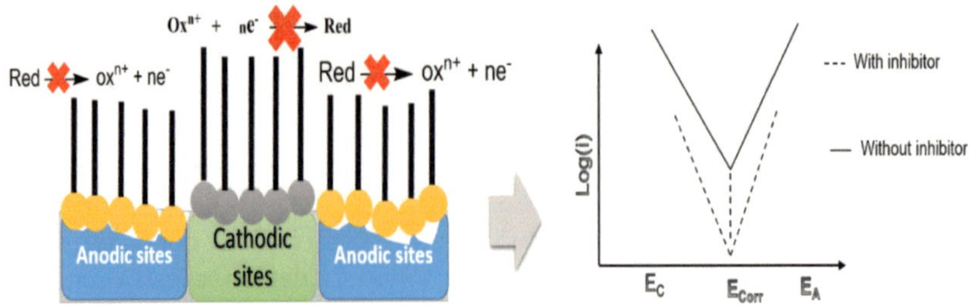

Figure 4.
Effect of addition of the mixed inhibitor.

Mixed inhibitors can decrease the cathodic and anodic reaction rates at the same time because they affect the oxidation and reduction reaction, with little change in the corrosion potential (less than 85 mV around the corrosion potential) [6] (**Figure 4**).

3.1.1 Action mode of the corrosion inhibitors in liquid phase

Each type of inhibitor is characterized by its action mode: adsorption, barrier, reinforcing of the oxide layer, passivation, and formed insoluble complex.

In the case of the interposition of a barrier between the metal and the corrosive medium, which is essential in acidic backgrounds, the role of adsorption of the compounds on the surface is essential.

The reinforcement of a pre-existing barrier, in general the oxide or hydroxide layer formed naturally in a neutral or alkaline medium, may consist of an extension of the oxide to the surface or of the precipitation of salts with weak places of the oxide, these salts being corrosive agents.

The formation of the barrier by interaction between the inhibitor and one or more species of the corrosive medium is a type of mechanism which is also specific for neutral or alkaline media.

Obviously, taking into these general notions, the mechanism of action of an inhibitor can be considered under two aspects: a mechanistic aspect (intervention in the fundamental corrosion processes) and a morphological aspect (intervention of the inhibitory molecule in the interfacial structure). It is also clear that the mechanism of action will differentiate strongly depending on the pH characteristics of the medium.

3.2 Gas phase: volatile corrosion inhibitors

Volatile corrosion inhibitor is referred to as gas molecules used as a corrosion inhibitor; they are intended for the temporary protection of metallic materials placed in the atmosphere, essentially in storage or transport condition. Their use is made either in the form of wrapping papers impregnated with product or in the form of powder or by spraying with a solution (volatile solvent) [8] (**Figure 5**).

Vapor phase inhibitors (VPI) or volatile corrosion inhibitors (VCI) are low nitrogen base salts (cyclohexylamine, dicyclohexylamine, guanidine), and weak acids (nitrous acid, carbonic acid, benzoic acid). The organic part ensures volatility and a certain protective power, and the inorganic part adjusts the volatility, which must correspond to vapor pressures between 10^{-4} and 10^{-2} mmHg at room temperature, and ensures the supply of groups of protectors (Ph-COO–...).

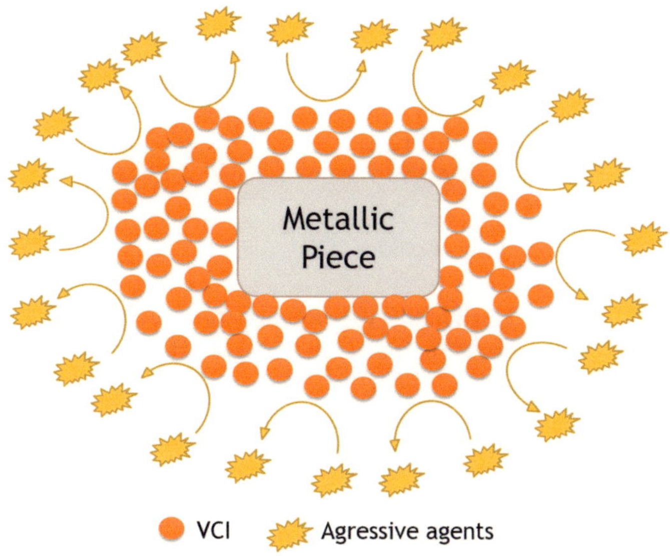

Figure 5.
Volatile corrosion inhibitor mechanism.

The inhibitor molecules acts by different ways; they are transported or dissociated on the metal surface to ensure an excellent protection by:

- pH effect
- Adsorption leading to hydrophobization of the surface
- Effect on electrochemical processes, essentially on the anodic process: blocking of sites or passivation

The adsorption is more of a chemical type, and the molecule is difficult to remove afterwards. Despite this, the protective action is only maintained if the source of the inhibitor is itself maintained in the immediate environment of the metal surface.

4. Particular inhibitor

The molecule inhibitors have three areas of application which are in particular important for the use of these products: the petroleum industry, water treatment, and pickling/cleaning of metals. Other applications exist for inhibitors, which involve then more specific formulations, which will be described in the following.

4.1 Inhibitors for coating

These are essentially the inhibitors added to the paints.

The incorporation of a pigment having inhibitory properties in the coating primer provides most of the corrosion resistance provided by the coating. There are two types of pigments: the active inhibitor pigment and the inactive inhibitor pigment.

The *active inhibitor pigments*, partially soluble in water, exist on the surface metal under the coating, providing active protection of the metal [9].

At the same time, it helps maintain an optimal ratio between the pigment volume concentration and the critical pigment volume concentration that means the film is not too compact to avoid blistering nor too loose to prevent the penetration of aggressive ions (Cl^-).

- Main inhibitors used
    - o Zinc-based formulations include:
    - o Zinc powder, which provides cathodic protection of the steel coated, provided that the dry extract is greater than 92% of the mass paint.
    - o Zinc chromate, used because of its solubility in water; very effective in paints on aluminum, as well as strontium chromate.
    - o Lead-based formulations include:
    - o Lead powder, whose mode of action is complex and passing probably by modifying the pH of the aqueous medium on contact paint (alkalization) and metal and reverse polarity between iron and lead, explaining cathodic protection of the ferrous material.
    - o Lead oxide Pb_3O_4, many mechanisms of action are proposed: formation of mixed protective layers of PbO oxide corrosion products, formation of soaps with constituents of paint, etc.

The *inactive inhibitor pigments* are essentially iron oxides, natural or synthetic, whose role is only to adjust the pigment volume concentration around the critical concentration.

5. Inhibitors for industrial section

Although their use could theoretically be envisaged in most cases of corrosion (with, as main limitations, too large volume of the corrosive medium or the possible impossibility of incorporating additives therein), inhibitors have several traditional application fields [10]:

- Water treatment (sanitary water, industrial process water, boiler water, etc.).
- Petroleum industry: drilling, extraction, refining, storage, and transport; at all stages of this industry, the use of corrosion inhibitors is essential for the safeguarding of installations.
- Temporary protection of metals, whether during acid pickling and cleaning of installations or storage in the atmosphere (volatile inhibitors, incorporation into oils and greases for temporary protection) or for the treatment of cutting oils.
- Paint industry on metals where the inhibitors are additives ensuring the anticorrosion protection of metals.

6. Green corrosion inhibitors (GCI)

GCI is the molecule extract from plants; it has a double effect because it has a good ability to protect the metals and no negative effect on the environment at the same time.

In the last few years, the researcher community had oriented to inhibitors extracted from plants such as essential and vegetable oil, flavonoids, coumarins, steroids, terpenoids, and condensed tannins. These substances are excellent inhibitors because they contain heteroatoms such as an N, O, P, and S. The free electrons on the heteroatoms form bonds with the electrons on the metal surface. Some atoms in water ionize to release a proton; thus, the now negatively charged heteroatom helps to free an electron on the heteroatom and forms a stronger bond with the metallic electrons. These properties confer them good inhibition properties.

The following table summarized some green inhibitors used for corrosion inhibition of steels, steel alloys, aluminums, aluminum alloys, and other metals and alloys (**Table 1**).

Metal	Inhibitor source	Active ingredient	References
Steel	Tamarind		[11]
Steel	Tea leaves		[12]
Steel	Eucalyptus oil	Monomtrene 1,8-cineole	[13]
C-steel, Ni, Zn	Lawsonia extract (henna)	Lawsone (2-hydroxy-1,4-napthoquinone resin and tannin, coumarine, Gallic, acid, and sterols)	[14]
Mild steel	Gum exudate	Hexuronic acid, neutral sugar residues, volatile monoterpenes, canaric and related triterpene acids, reducing and nonreducing sugars	[15]
Mild steel	Garcinia kola seed	Primary and secondary amines Unsaturated fatty acids and biflavnone	[16]
Steel	Aloe leaves		[17]
Steel	Mango/orange peels		[18]
Steel	Hibiscus sabdariffa (Calyx extract) in 1 M H_2SO_4 and 2 M HCl solutions, stock 10–50%	Molecular protonated organic species in the extract. Ascorbic acid, amino acids, flavonoids, Pigments and carotene	[19]
Al-Mg alloy	Aqueous extract of *Rosmarinus officinalis*—neutral phenol subfraction of the aqueous extract	Catechin	[20]
Al	Opuntia (modified stems cladodes)	Polysaccharide (mucilage and pectin)	[21]
Zn	Metal chelates of citric acid		[22]
Zn	Onion juice	S-containing acids (glutamyl peptides) S-(1-propenyl) L-cysteine sulfoxide, and S-2-carboxypropyl glutathione	[23]
Sn	Natural honey (acacia chestnut)		[24]
Sn	Black radish		[25]

Table 1.
Green inhibitors used for corrosion inhibition of steel, steel alloys, aluminum, aluminum alloys, and other metals and alloys.

7. Green corrosion inhibitor study

In this part, we will present a study of a green corrosion inhibitor using a formulation prepared based on the *Ceratonia siliqua* L. seed oil noted FCSL in simulated acid rain solution with pH equal to 3.6. This medium is simulated to acid rain in urban zone [26].

The inhibition effect was evaluated using the electrochemical measurement such as polarization curves and spectroscopy impedance. The electrode surface was characterized by SEM/EDS.

7.1 Open circuit potential

The results of the open circuit potential (OCP) variation of the iron substrate in acidic solution in the presence and in the absence of FCSL are reported in **Figure 6**.

The results show that in the absence of FCSL, the potential tends to stabilize at −0.51 V, after 20 min.

The addition of the FCSL formulation leads to a shift in the corrosion potential to a positive direction. This important shift of corrosion potential may indicate an important anodic inhibiting effect of FCSL.

7.2 Potentiodynamic curves

The polarization studies of iron were carried out in acid rain solution both in the absence and in the presence of the FCSL formulation.

All of these curves were obtained after 30 min of immersion time of the electrode in electrolytic solution after performing the automatic ohmic drop compensation (ZIR).

The cathodic and anodic polarization curves of iron in simulated acid rain solution with and without various inhibition concentrations are reported in **Figure 7**.

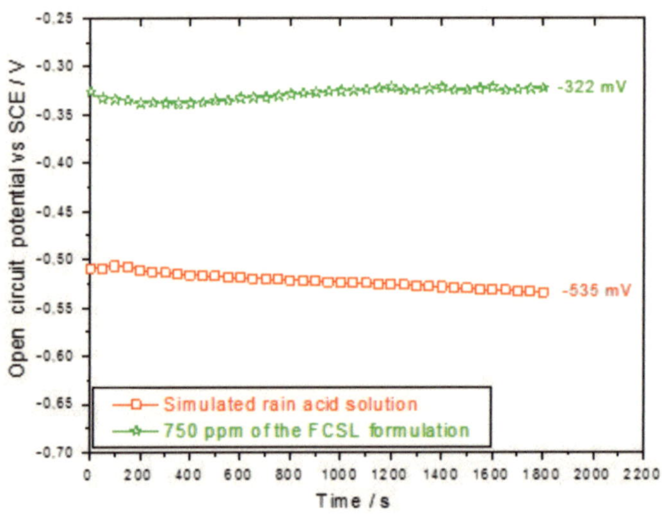

Figure 6.
Variation of the open circuit potential (OCP) of the iron substrate in acidic solution with and without the FCSL formulation.

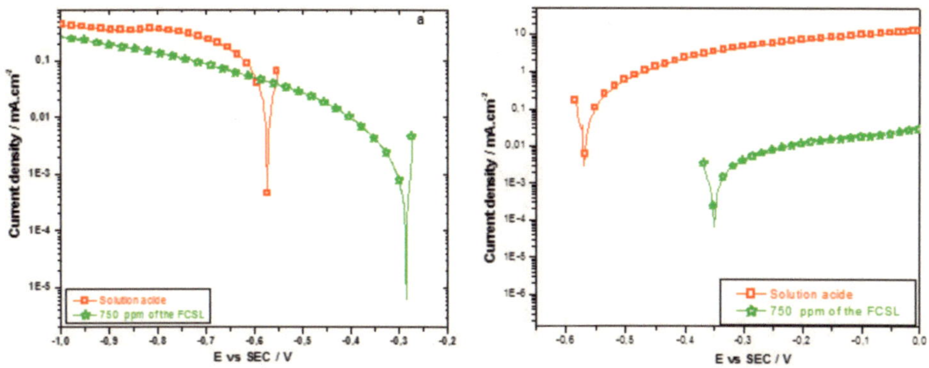

Figure 7.
Potentiodynamic polarization curves of the iron in acid rain solution with and without various inhibition concentrations in the cathodic domain (a) and in the anodic domain (b).

7.2.1 The cathodic polarization

In absence of the inhibitor (**Figure 7a**), the corrosion current increases rapidly with the cathodic overvoltage until the potential value of −0.8 V/SCE; for more negative potential values, a pseudo-plateau appears in 0.4 mA/cm^2, which can be attributed to the oxygen diffusion process, so the cathodic reaction can be expressed in the following equation:

$$2H_2O + O_2 + 4e^- \rightarrow 4OH^- \quad (1)$$

In the cathodic process, the important factor that must be considered is the mass transport [24].

The adding of the formulation to the corrosive solution is accompanied by both a shift of corrosion potential E_{corr} toward a more positive potential and a decrease of the current density I_{corr}, with the disappearance of the diffusion plateau, obtained in the case of the blank solution, in the presence of the FCSL formulation. We observe the formation of the film on the area, which hinders the diffusion of dissolved oxygen toward the electrode surface.

7.2.2 The anodic polarization

$$Fe \rightarrow Fe^{2+} + 2e^- \quad (2)$$

This behavior is associated with the presence of chloride in solution. The anodic reaction can be expressed by the following equation:

According to **Figure 7b** in the presence of the FCSL formulation, the current density decreases significantly with the presence of the inhibitor. Furthermore, the corrosion potential displayed more positive values, and also the value of the corrosion current density I_{corr} reduced from 74.9 mA cm^{-2} in the absence of the inhibitor to 1.0 mA cm^{-2} in the presence of the inhibitor. So in less than 75 times, then we observe the appearance of a current plateau in a wide potential range.

The inhibition efficiency (% IE) was calculated using the following relation and its value of 98.6%:

$$E(\%) = \frac{i_0^{corr} - i_0^{Inh}}{i_0^{corr}} \times 100 \quad (3)$$

The corresponding current plateau value is in the order of 0.03 mA cm^{-2} in the case of the FSCL formulation. This may indicate that the iron surface is protected by the inhibitor; this protection may be attributed to a passivity of iron substrate resulting from the formation of inhibitor film on the iron electrode surface. Thus results were observed by other authors [27].

From this result, we can conclude that the FSCL formulation is a mixed-type inhibitor that acts by decreasing the current density in both the cathodic and the anodic domains and making the corrosion potential become more anodic.

This good inhibiting effect of the FSCL formulation may be related to the adsorption on the electrode surface by the establishment of a barrier film.

7.3 Electrochemical impedance spectroscopy

The impedance diagrams in Nyquist and Bode plots in the absence and in the presence of the FCSL at 293 K are represented in **Figure 8**.

In the case of the blank solution, as shown in **Figure 8**, we noted the two capacity loops in the high frequencies and the inductive loop at low frequencies. This inductive effect may be due to the desorption of the H^+ ions and salt ions present in the solution or to the redissolution of the passivity surface [26]. In effect, this inductive loop disappeared with the addition of the inhibitor. The same behavior has been observed by other authors [27].

As shown in **Figure 8**, in the presence of the FCSL formulation, the size of the loops are bigger than in the case of the blank. Indeed, the polarization resistances pass from 380 Ω cm^2 in the case of blank to 14,080 Ω cm^2 in the presence of inhibitor.

The inhibition efficiency (% IE) was calculated using the following relation and its value of 97.3%:

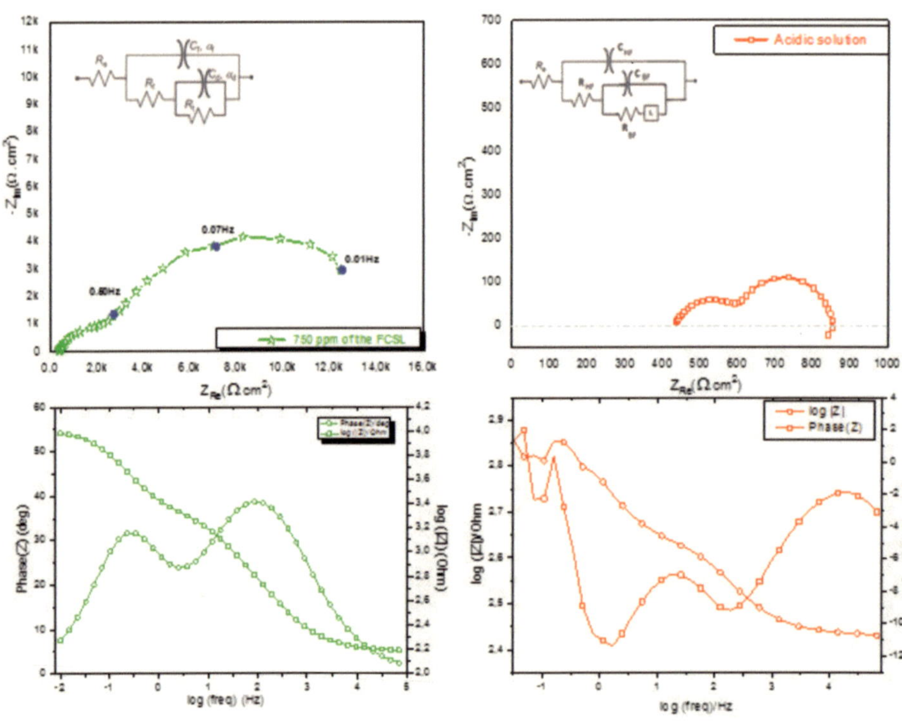

Figure 8.
Nyquist and Bode impedance plots of the iron electrode in acidic solution with and without the FCSL.

$$E(\%) = \frac{R_P^{corr} - R_P^{Inh}}{R_P^{corr}} \times 100 \quad (4)$$

We noted also a decrease in the electrolyte resistance which may be explicated by the presence of the ionized substances in the formulation.

We can be ascribed to the following contributions. The high-frequency contribution (C_f, R_f) can be attributed to the dielectric character due to the formation of the film on the iron surface in presence of the inhibitor. The low-frequency contribution can be attributed to the double-layer capacitance (C_{dl}) at the electrolyte/iron interface at the bottom of the pores coupled with the charge transfer resistance (R_t) [10].

7.4 Surface analysis

The aim of the surface analysis by SEM coupled with the EDX used in this study is firstly to check the hypothesis of the formation of the inhibitor on the electrode surface and secondly to verify its protective qualities against iron corrosion.

Figures 9 and **10** show the area of the substrate of iron with and without the FCSL formulation after 24 h of immersion time in the acidic solution.

In the absence of the FSCL formulation (**Figure 9**), the SEM examination provides that the metallic surface be heavily attacked by the corrosive ions. The EDX spectrum reported in **Figure 9** showed the characteristic peaks of the specimen and marked the presence of oxygen atoms.

Also in the case of the presence the FSCL formulation (**Figure 10**), in the addition of the optimal concentration of the FSCL formulation into the corrosion

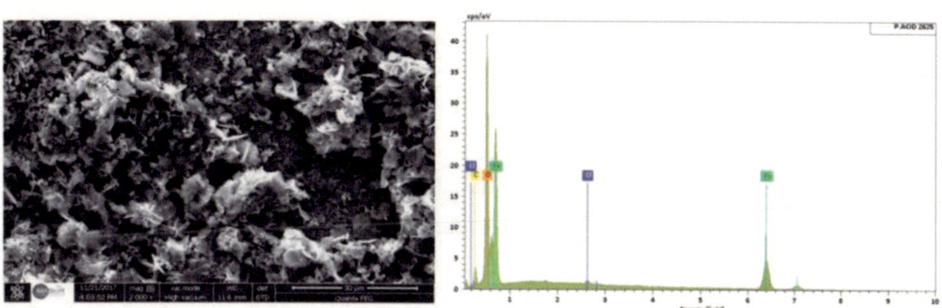

Figure 9.
SEM micrograph and EDS spectrum of the iron substrate in corrosive solution after 24 h of immersion time in the absence of the FCSL formulation.

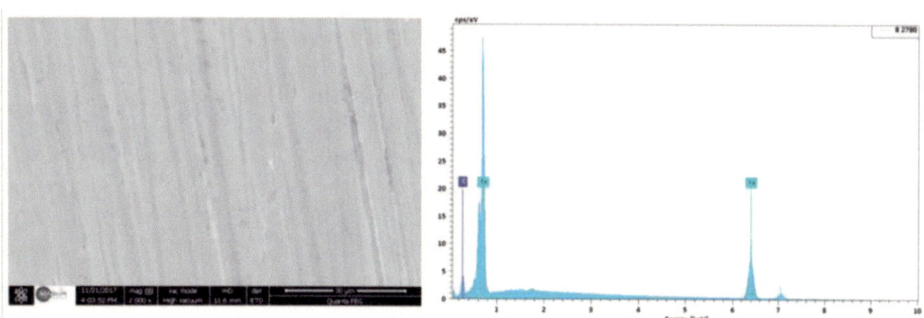

Figure 10.
SEM micrograph and EDX spectrum of the iron substrate in acidic solution after 24 h of immersion time in the presence of the FCSL formulation.

solution, a smooth surface noted could explain the good protection effect of the inhibitor by a formation of the film. As confirmed by the EDX spectrum, a very low content of oxygen species is revealed.

8. Conclusion

Corrosion is one of the most destructive phenomena that can affect metallic pieces. Through this work, we present one of the most used method to protect the metals: the corrosion inhibitor. In addition, we present a case study using a green corrosion inhibitor prepared from the oil of *Ceratonia siliqua* L. seeds.

Declaration of competing interest

The authors have declared that no conflict of interest exists.

Author details

Said Abbout
University Ibn Tofail, Kenitra, Morocco

*Address all correspondence to: said.abbout@uit.ac.ma; said_about@hotmail.fr

IntechOpen

© 2020 The Author(s). Licensee IntechOpen. This chapter is distributed under the terms of the Creative Commons Attribution License (http://creativecommons.org/licenses/by/3.0), which permits unrestricted use, distribution, and reproduction in any medium, provided the original work is properly cited.

References

[1] Yusoff MFM, Afiqah MI, Zulkafli Y, Othman NK, Lazim A, Mokhtar WNAW. Temperature effects toward corrosion rate of carbon and mild steel using red palm oil as natural corrosion inhibitor. Malaysian Journal of Chemistry (MJCHEM). 2019;**21**:64-70

[2] ISO 8044:2015 (en). Corrosion of metals and alloys—Basic terms and definitions

[3] Zhang F, Chen C, Hou R, Li J, Cao Y, Dong S, et al. Investigation and application of mussel adhesive protein nanocomposite film-forming inhibitor for reinforced concrete engineering. Corrosion Science. 2019;**153**:333-340

[4] Mitachi K, Kurosu SM, Eslamimehr S, Lemieux MR, Ishizaki Y, Clemons WM Jr, et al. Semisynthesis of an anticancer DPAGT1 inhibitor from a muraymycin biosynthetic intermediate. Organic Letters. 2019;**21**:876-879

[5] Gangadhara G, Dahl G, Bohnacker T, Rae R, Gunnarsson J, Blaho S, et al. A class of highly selective inhibitors bind to an active state of PI3Kγ. Nature Chemical Biology. 2019;**15**:348

[6] Tang Z. A review of corrosion inhibitors for rust preventative fluids. Current Opinion in Solid State and Materials Science. 2019;**23**:100759

[7] Frolenkova S, Overchenko T, Motronyuk T, Vorobyova V, Miroshnychenko I, Panchenko M. Passivating anions effect on the anodic behavior of steel in a converting acetate solution. Journal of Chemical Technology and Metallurgy. 2019;**54**:443-446

[8] Vigdorovich V, Tsygankova L, Knyazeva L. Universality of volatile corrosion inhibitors in terms of agricultural production requirements. Russian Agricultural Sciences. 2019;**45**:307-311

[9] Palanisamy G. Corrosion Inhibitors. In: Corrosion Inhibitors. IntechOpen; 2019

[10] Umoren SA, Solomon MM, Obot IB, Sulieman RK. A critical review on the recent studies on plant biomaterials as corrosion inhibitors for industrial metals. Journal of Industrial and Engineering Chemistry. 2019

[11] Henriquez-Román J, Sancy M, Páez M, Padilla-Campos L, Zagal J, Rangel C, et al. The influence of aniline and its derivatives on the corrosion behaviour of copper in acid solution. Journal of Solid State Electrochemistry. 2005;**9**:504-511

[12] El Hosary A, Saleh R, El Din AS. Corrosion inhibition by naturally occurring substances—I. The effect of Hibiscus subdariffa (karkade) extract on the dissolution of Al and Zn. Corrosion Science. 1972;**12**:897-904

[13] Bouyanzer A, Majidi L, Hammouti B. Effect of eucalyptus oil on the corrosion of steel in 1M HCl. Bulletin of Electrochemistry. 2006;**22**:321-324

[14] Chetouani A, Hammouti B. Corrosion inhibition of iron in hydrochloric acid solutions by naturally henna. Bulletin of Electrochemistry. 2003;**19**:23-25

[15] Umoren S, Obot I, Ebenso E. Corrosion inhibition of aluminium using exudate gum from Pachylobus edulis in the presence of halide ions in HCl. Journal of Chemistry. 2008;**5**:355-364

[16] Okafor P, Osabor V, Ebenso E. Eco-friendly corrosion inhibitors: Inhibitive action of ethanol extracts of *Garcinia*

kola for the corrosion of mild steel in H_2SO_4 solutions. Pigment & Resin Technology. 2007;**36**:299-305

[17] Ramachandra C, Rao PS. Processing of Aloe vera leaf gel: A review. American Journal of Agricultural and Biological Sciences. 2008;**3**:502-510

[18] Ezugwu A, Ezike T, Ibeawuchi A, Nsude C, Udenwobele D, Eze S, et al. Comparative studies on pectinases obtained from *Aspergillus fumigatus* and *Aspergillus niger* in submerged fermentation system using pectin extracted from mango, orange and pineapple peels as carbon sources. Nigerian Journal of Biotechnology. 2014;**28**:26-34

[19] Oguzie EE. Corrosion inhibitive effect and adsorption behaviour of *Hibiscus sabdariffa* extract on mild steel in acidic media. Portugaliae Electrochimica Acta. 2008;**26**:303-314

[20] Kliškić M, Radošević J, Gudić S, Katalinić V. Aqueous extract of *Rosmarinus officinalis* L. as inhibitor of Al–Mg alloy corrosion in chloride solution. Journal of Applied Electrochemistry. 2000;**30**:823-830

[21] El-Etre A. Inhibition of aluminum corrosion using Opuntia extract. Corrosion Science. 2003;**45**:2485-2495

[22] Müller B. Amino and polyamino acids as corrosion inhibitors for aluminium and zinc pigments. Pigment & Resin Technology. 2002;**31**:84-87

[23] El-Etre A. Natural onion juice as inhibitor for zinc corrosion. Bulletin of Electrochemistry. 2006;**22**:75-80

[24] Vrsalović L, Gudić S, Kliškić M. *Salvia officinalis* L. honey as corrosion inhibitor for CuNiFe alloy in sodium chloride solution. Indian Journal of Chemical Technology. 2012;**19**(2):96-102

[25] Left DB, Zertoubi M, Irhzo A, Azzi M. Revue: Huiles et Extraits de plantes comme inhibiteurs de corrosion pour différents métaux et alliages dans le milieu acide chlorhydrique. [Review: Oils and extracts plants as corrosion inhibitors for different metals and alloys in hydrochloric acid medium]. Journal of Materials and Environmental Science. 2013;**4**(6):855-866

[26] Abbout S, Chellouli M, Zouarhi M, Benzidia B, Hammouch H, Chebabe D, et al. New formulation based on *Ceratonia siliqua* L seed oil, as a green corrosion inhibitor of iron in acidic medium. Analytical & Bioanalytical Electrochemistry. 2018;**10**:789-804

[27] Zouarhi M, Chellouli M, Abbout S, Hammouch H, Dermaj A, Said Hassane SO, et al. Inhibiting effect of a green corrosion inhibitor containing *Jatropha curcas* seeds oil for iron in an acidic medium. Portugaliae Electrochimica Acta. 2018;**36**:179-195

Chapter 3

Introduction of Inhibitors, Mechanism and Application for Protection of Steel Reinforcement Corrosion in Concrete

Anil Kumar

Abstract

The corrosion of steel reinforcement in concrete due to environmental factors has been studied through numerous approaches and the reduction of corrosion has been managed by various methods; however, among the protection techniques, the use of corrosion inhibitors has gained encouragement. In this chapter, nitrites and nitrates of sodium and calcium and sodium molybdates and sodium tungstates (oxyanions of group VI) were studied and have gained sufficient scientific coverage. However, their exact role of inhibition was studied by simple polarization technique. In this chapter, we compare the inhibitive efficiency of nitrites and nitrates of sodium and calcium and also that of molybdates and tungstates. The results, however, indicate that among nitrites and nitrates, the calcium salts are more efficient and molybdates and tungstates are comparable in their inhibitive efficacy.

Keywords: corrosion, steel reinforcement, inhibitors, polarization, mechanism, I_{corr}

1. Introduction

Corrosion is the destructive attack upon a metal by its environment and it is an electrochemical phenomenon. Most common examples of corrosion include the rusting of iron and steel, tarnishing of silver and copper, blistering and bubbling of chromium plating and paintwork on cars, discharge of rust-coloured water from domestic taps and seizure of nuts and bolts. Corrosion leads to weakening of metal structures, failure of plant and pollution of process liquors. It is necessary to understand the basic principles of corrosion before taking appropriate preventative or protective measures. The various efforts towards reducing corrosion of metals can be grouped into the following: modification of bulk alloys, modification of environments and surface modifications. Here, in this chapter, we discuss the modification of environments by adding small concentration of inhibitors. An inhibitor is a chemical substance that inhibits or effectively decreases the corrosion rate. It can be understand by **Figure 1**, I_{corr}. decreases with inhibitors in comparison to without inhibitors.

IntechOpen

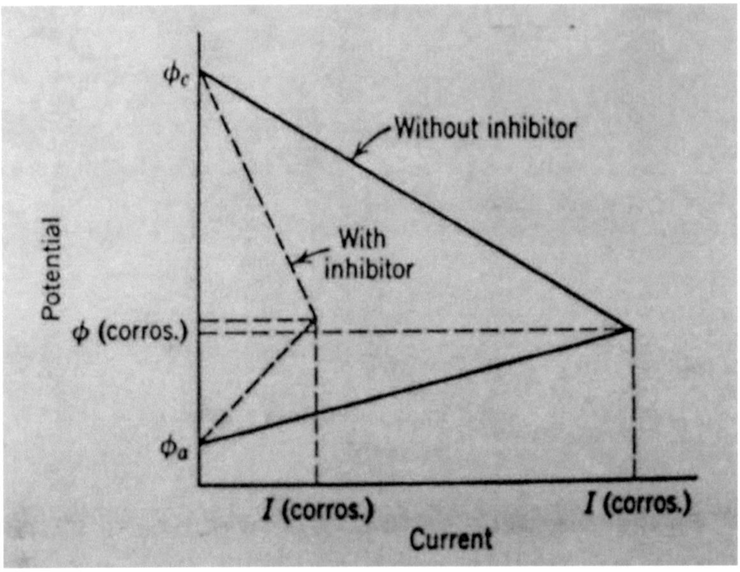

Figure 1.
Polarization diagram for steel corroding in pickling acid with and without inhibitor.

2. Types of inhibitors

There are several classes of inhibitors designated as follows:

2.1 Passivators

Passivators are usually inorganic oxidizing substances that passivate the metal and shift the corrosion potential several tenths of a volts in the noble direction, in which the metal reacts with inhibitors to form inhibiting layer or film on the metal surface. The following inhibitors act as passivators: chromates, nitrates, molybdates etc.

2.2 Non-passivating inhibitors

These include inhibitors such as the pickling inhibitors, which are usually organic substances that have only slight effect on the corrosion potential, changing it either in the noble or active direction usually by not more than a few milli or centivolts. They form an absorbed monolayer thickness on the metal surface, which essentially blocks the discharge of H^+ and dissolution of metals ions. Some inhibitors block the cathodic reaction (raise hydrogen overvoltage) more than the anodic reaction or vice versa; but adsorption appears to be generally over the whole surface rather than at specific anodic or cathodic sites, and both reactions tend to be retarded. Hence, on addition of an inhibitor to an acid, the corrosion potential of steel is not greatly altered (<0.1 V), although the corrosion rate may be appreciably reduced as shown in **Figure 1**. Generally, polar group compounds serve as pickling inhibitors. These include the organic nitrogen, amine, S and OH groups' compound. The size, orientation, shape and electric charge of the molecule play a part in the effectiveness of inhibition. Whether a compound adsorbs on a given metal and the relative strength of the adsorbed bond often depend on factors such as surface charge of metal. For inhibitors that adsorb better at increasingly active potentials as measured from the point of so-called zero surface charge (potential of minimum ionic adsorption), cathodic polarization in the presence of the inhibitor provides better protection than either the equivalent cathodic protection or use of an inhibitor alone. This was demonstrated by Antropov for iron and zinc in sulphuric acid containing various organic inhibitors. In general the passive-type inhibitors reduce corrosion rate to very low values, being more efficient in this regard than most of the non-passivating types. They represent, therefore, the best inhibitors available for certain metal-environment combinations.

2.3 Slushing compounds inhibitors

Slushing compounds are used to protect steel surface temporarily from rusting during shipment or storage. They consist of oil, greases or waxes that contain small amounts of organic additives. The additives are polar compounds that absorb on the metal surface in the form of a closely packed oriented layer. Organic amines, zinc naphthenate, various oxidation products of petroleum, alkali and alkaline earth metal salts of sulphonate oils, and various other compounds were used as additive for slushing compound.

2.4 Vapour-phase inhibitors

Vapour phase inhibitors are basically soluble film-forming inhibitors, such as nitrites, benzoates or carbonates, which are attached to a large organic cation. The large organic cation functions as a parachute or umbrella for the organic cation and the inorganic anion will provide the molecule with a vapour pressure that will lie within the range of 0.1–1.0 mm mercury (at ambient temperatures). Through this, the inhibitor will slowly evaporate and provide an adequate supply of inhibitor at the metal surface for protection if required, but the evaporation rate will not be so fast that the inhibitor is lost in too short a time. It moisture condenses on the metal surface, the vapour phase inhibitor will dissolve and the concentration of inhibitive anions will be sufficient to ensure passivation of the metal. It is possible, however, that there is conjoint action between the passivating anion inhibitor and some form of adsorption inhibition by the organic cation, the latter being assisted by anions such as chloride or sulphate shifting the corrosion potential to move negative values. They are used to protect critical machine parts (e.g., ball bearing or other manufactured steel articles) temporarily against rusting by moisture during shipping or storage. It appears to be one of adsorbed film formation on the metal surface that provides protection against water or oxygen, or both. In the case of volatile nitrites, the inhibitor may also supply a certain amount of NO_2^-, which passivates the surface. Examples of vapour-phase inhibitors are dicyclohexylammonium nitrite, cyclohexylamine and ethanolamine carbonate. Dicyclohexylammonium nitrite has been one of the most effective of the vapour-phase inhibitors for corrosion of steel for a period of years. This substance is white crystalline, almost odorless and relatively non-toxic. It has a vapour pressure of 0.0001 mm of Hg at 21° C (70° F), which is about one-tenth the vapour pressure of mercury itself. One gram saturates about 550 m^3 of air, rendering the air relatively noncorrosive to steel. However, it should be used with caution in contact with non-ferrous metals. In particular, corrosion of zinc, magnesium and cadmium is accelerated. Cyclohexylamine carbonate has a somewhat higher vapour pressure of 0.4 mm Hg at 25°C and its vapour also effectively inhibits steel. The higher vapour pressure provides more rapid inhibition of steel surfaces either during packaging or on opening and again closing a package, during which time concentration of vapour may fall below that required for protection. The vapour reduces the corrosion of aluminium, solder and zinc but it has no inhibition effect on cadmium and it increases corrosion of copper, brass and magnesium. Copper dipped in 0.25%

benzotriazole in water at 60°C for 2 min forms a thin adsorbed film that protects the metal against subsequent tarnish in the atmosphere, which acts as a diffusion barrier known as inhibitor to tarnishing of copper. It is also effective for brasses and for nickel-silver (zinc-nickel alloys).

The overall inhibitors may be classified on their nature as follows:

1. Type A: Substances that form an inhibiting layer or film on the metal, that is passivation of metal. Type A inhibitors may be subdivided as follows:

    i. Type IA: Inhibitors that that reduce the corrosion rate but do not completely prevent corrosion, for example pickling inhibitors (**Figure 2**).

    ii. Type IIA: Inhibitors that delay the onset of corrosion for long periods so that the metal has a temporary immunity to corrosion, for example, benzotriazole for copper (**Figure 3**).

    iii. Type IIIA: Passivating inhibitors, which result in the formation of passive films on the metal surface, the films generally being an oxide or an insoluble salt of the metal, for example, NaNO2, phosphates and chromates for steel, sulphates for lead (**Figure 4**).

2. Type B: Substances that reduce the aggressiveness of the environment, which may be subdivided as follows:

    i. Type IB: Inhibitors that retard the corrosion process but do not completely prevent it.

    ii. Type IIB: Inhibitors that incubate corrosion by reacting with substances that produce corrosion in a given environment, for example, organic amines, which 'neutralize' organic acid in oils and emulsion. In another way, it may be divided into two types of corrosion inhibitors:

        a. Organic inhibitors: Inhibitors of Type IA, IIA and IIB are organic compounds and

        b. Inorganic inhibitors: Inhibitors of Type IIIA and IB are inorganic compounds.

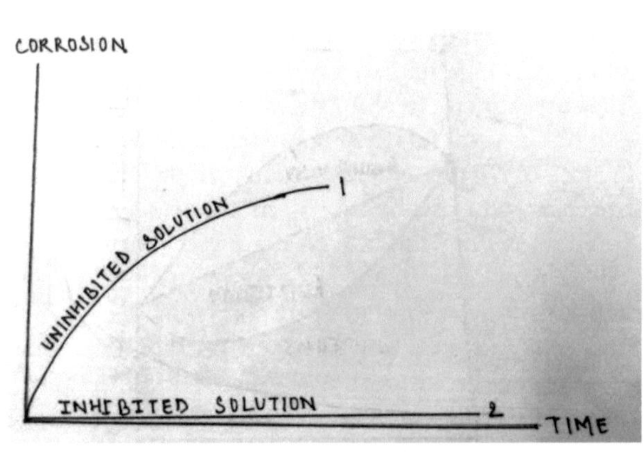

Figure 2.
Effect of type IA inhibitors on corrosion.

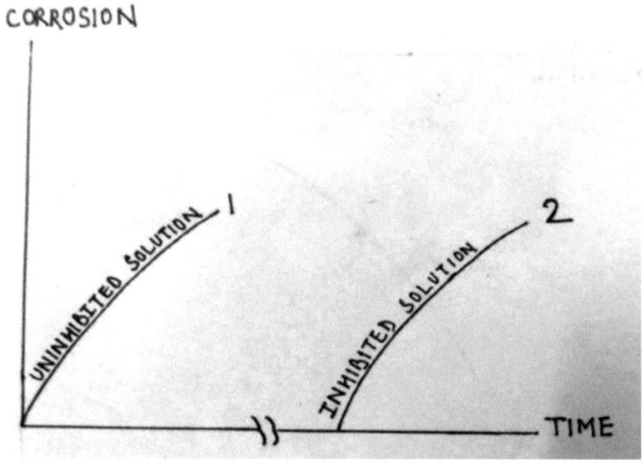

Figure 3.
Incubation of corrosion by type IIA inhibitors.

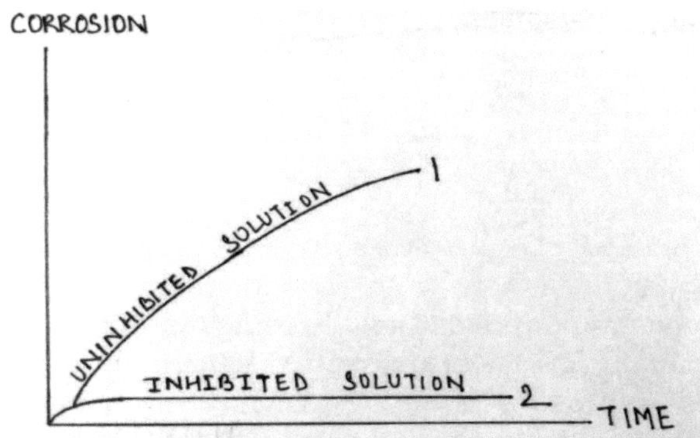

Figure 4.
Effect of passivating type IIIA inhibitors on corrosion.

2.5 Mechanism of inhibitors

Inhibitors act as depolarizers, when metal surface comes in contact with passivators. In this case, they initiate high current densities at residual anodic areas, which exceed $i_{critical}$ for passivation as shown in **Figure 2**. Generally, those substances are oxidizing agents with redox potentials that are nobler than that of the metal and they are readily reduced, resulting in a very high cathodic current density and consequently a high anodic current density, which is greater than the critical current density. Consequently, the metal will passivate after a short initial period of dissolution. Ions such as sulphates and chlorides do not passivate iron since they are not easily reduced and nitrates are poor passivating agents since they, too, are only sluggishly reduced (**Figure 2**).

Only those ions can serve as passivators that have both an oxidizing capacity in the thermodynamic sensor (noble oxidation reduction potential) and that are readily reduced with shallow cathodic polarization curve (**Figure 5**). Passivating agents are reduced at cathodic areas at a current density equivalent to that at the anodic areas, which is itself greater than the critical current density of the metal such that passivation of the metal occurs. For the optimum inhibition of corrosion, that is complete passivation of the metal surface, the inhibitor concentration must be greater than a critical concentration. This critical concentration varies with the type of inhibitor but it is generally within the range of 10^{-4} to 10^{-3} M; but it will

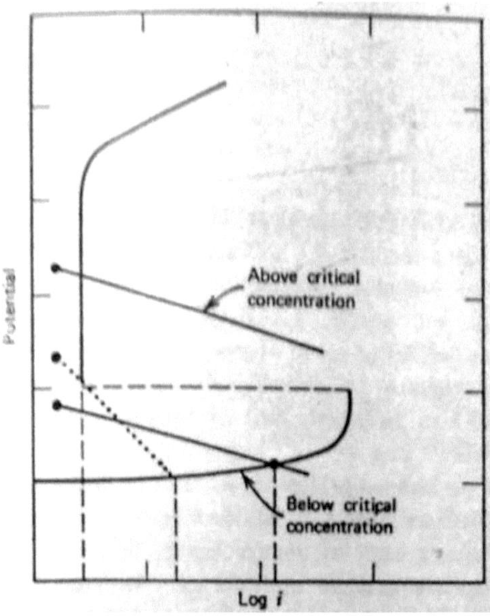

Figure 5.
Polarization curves that show the effect of passivator concentration on corrosion of iron. An oxidizing substance that reduces sluggishly does not induce passivity (dotted cathodic polarization curve).

be greater in the presence of high hydrogen ion or chloride ion concentrations and at elevation temperatures. If the inhibitor concentration lies below the critical level, incomplete passivation is found and the passivator can function as an active depolarizer at other (unpassivated) areas such that increased corrosion rates are observed in localized areas, that is, pitting occurs. Thus, as the inhibitor falls below the critical concentration, either through consumption of the inhibitor or through inadequate or though inadequate dosing in stagnant areas, the more active redox potential of the inhibitor results in cathodic polarization curves that intersect the anodic polarization curve within the active region (**Figure 5**). Generally, the cathodic areas are larger than the anodic areas so that passivity becomes established on the metal surface. The passive areas become noble to adjacent areas and passivity spreads over the entire metal surface. Once passivity of the entire metal surface is established, the whole surface functions as a cathode and further reduction of the passivator is very slow and only a small or residual passive current flow is found. This low current flow is that required for passivator reduction corresponding to slow chemical dissolution of the passive film by its environment. Indirect passivators or the so-called cathodic inhibitors have a different mechanism for inhibition. These substances, for example, sodium hydroxide, phosphate, silicate or borate, result in passivation of metals such as iron by facilitating the adsorption of oxygen. With these inhibitors, there is both anion adsorption and reaction with the inhibitor such that a film covering the metal surface is formed. In general, the corrosion (or open-circuit) potential falls to more active or base values (as opposed to the more noble E_{corr} values found with true passivators). These substances are, however, less efficient as inhibitors and the protective power is rarely greater than 80–90%. Hence, there is no single mechanism of inhibitors. Passivators form film on the metal surface or selectively adsorbed on to a active anodic or cathodic sites on the surface. This adsorption, even in the absence of metal-inhibitor chemical interactions, results in polarization of the anodic and/or cathodic reaction so that corrosion is retarded or inhibited. Benzotriazole inhibitors incubate corrosion by forming a chelate-type of reaction product with the copper metal. Then corrosion is retarded or even prevented as long as the chelate was present on the metal. Pickling inhibitors are those that are adsorbed on to the metal and ultimately form a layer of chemisorbed material by interaction with the metal. The effectiveness of inhibitor depends on the nature of the film, rather than its thickness. Non-specific adsorption of ions or molecules that can form ions, on a metal surface, is dependent upon the surface charge on the metal. At the point of zero charge (ZPC), which for mercury is the point at which it exhibits maximum surface tension or shows an electro capillary maximum, the metal is uncharged. Clearly, at the ZPC, adsorption of both ions and molecules on to the metal surface can occur. When adsorption occurs, the electro capillary maximum is reduced or depressed and ions or molecules that are absorbed and depress the capillary are known as capillary-active agents. Capillary-active anions are adsorbed at potentials positive to (i.e., cathodic to) the ZPC while cations are adsorbed at potentials negative to (or anodic to) the ZPC. When such adsorption occurs, the ZPC is shifted, in the case of anions, to slightly more negative values, and the potential shift is known as a theta potential. For inhibition by anions to occur, the potential of a metal surface must be held positive to the ZPC, that is the metal is positively charged, and this generally occurs during corrosion of the metal in acid solutions. In neutral or basic media, an additional agent such as oxygen is generally required to maintain the metal corrosion potential positive to the ZPC. Organic amines and heterocyclic compounds will be able to absorb on to the corroding iron surface in hydrochloric acid solutions since the corrosion potential is positive or cathodic to the ZPC. Steric effects, for example, the size and shape of the molecule forming the anion as well as the shift in ZPC produced by the absorption process, will determine the effectiveness of the inhibitor on the corrosion rate. Cation-forming molecules, for example, many organic sulphur compounds, as well as the surface-active or wetting agent will have no effect on corrosion because they are not adsorbed at potentials cathodic to ZPC. In sulphuric acid solutions, however, sulphur compounds are effective inhibitors because iron becomes negatively charged in H_2SO_4, that is, its corrosion potential is negative (on anodic) to ZPC. Many inhibitive anions, even those that 'passivate' the metal surface, must be absorbed on to the metal surface before metal passivation can occur. Benzoates will not passivate iron unless the corrosion potential is cathodic to ZPC; for this to occur, oxygen must be present in solution and, in fact, benzoates are poor inhibitors for iron in deaerated solution. In alkaline solutions, the presence of OH ions (pH 6.5 or higher) assists adsorption of benzoate and it is possible that the so-called synergistic effect of nitrite ions with benzoate ions with may be due to the capillary-active behavior of the nitrile ion. Benzoates do not inhibit iron corrosion in acid solutions since the cathodic reaction is predominantly hydrogen evolution and no hydroxyl ions are released at the cathodic sites to assist benzoate adsorption.

2.6 Effect of inhibitor concentration

In general, corrosion rates decrease by increase in inhibitor concentration. Their plot is a mirror image of adsorption isotherms for the most organic inhibitors. It is shown in **Figure 6**. But in specific conditions, there appears to be a limiting corrosion rate even in the presence of large amount of inhibitor (**Figure 7**). This effect is most common with pickling inhibitors and the increase in corrosion rate often occurs at levels above 1–2% inhibitor. The cause of this phenomenon may be is ascribed to the ability of many organic substances, when present at relatively high concentration in solution, to function as hydrogen acceptors and so depolarize the cathodic reaction, for example, aliphatic aldehydes in acid solution. This increase in attack is found with substances that cannot accept hydrogen, that is, there is no overt depolarizing action.

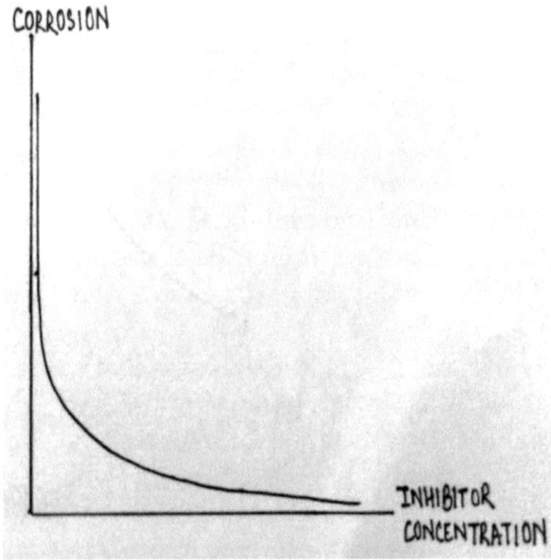

Figure 6.
Effect of inhibitor conc^n. On corrosion.

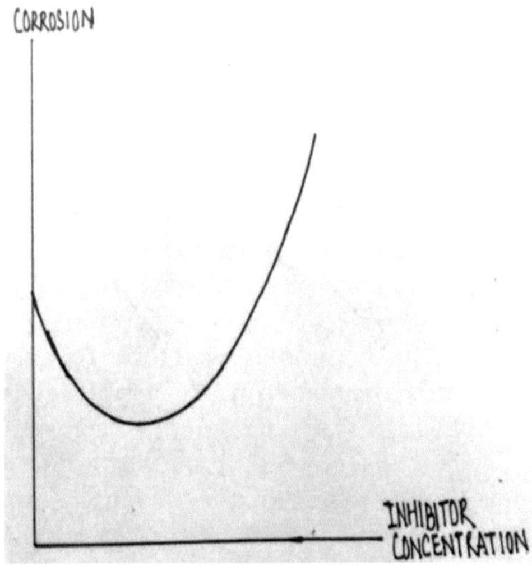

Figure 7.
Effect of depolarizing inhibitors on corrosion.

When two inhibitors are present in solution, it is rare that there is a simple additive effect of the combined protection power. In fact, three different effects can be observed when inhibitors are mixed (**Figure 8**).

Additive effect was found with similar structures and chemical properties of substance, for example, formaldehyde and acetaldehyde when used for steel in acid solution. It should also be noted that only very rarely do two inhibitors have the same retarding efficiency, R, for the same corrosion reaction.

$$\text{Retardation efficiency (or coefficient) } R = P_0/P.$$

$$\text{And protective power } P = P_0 - P/P \times 100\,\%.$$

where P and P_0 are the corrosion rates of the metal in the given medium with and without inhibitor respectively.

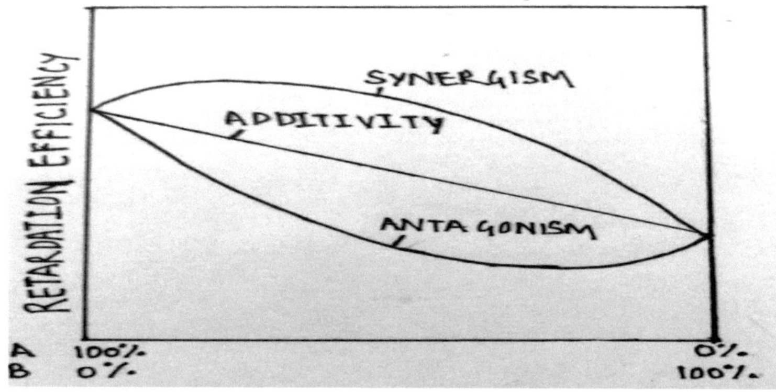

Figure 8.
Effect of mixtures of inhibitors on the retardation efficiency R of a corrosion reaction.

$$\text{Clearly, } R = 100/100 - P.$$

Synergism is the term applied to the marked reinforcement of the inhibiting action of one inhibitor by the addition of small amount of a second inhibitor, even though the second inhibitor is less effective when used alone. The retardation efficiency of the combination of inhibitors is considerably greater than a simple additive effect of the separate R value. An example is the effect of small addition of furfuralimine to formaldehyde for steel in sulphuric acid. The third effect that may be found for mixtures of inhibitors is antagonism or the mutual weakening of the inhibitive efficiency so that the retardation efficiency is lower (i.e., the corrosion rate is greater) for the mixture than for either substance alone. This effect is most common when there appears to be some form of chemical interaction between the two inhibitors. A typical example is the reduced efficiency of inhibition of steel in hydrochloric acid in the combined presence of antinomy chloride and aniline.

3. Application of inhibitors

3.1 Role of sodium nitrite inhibitor in rebar corrosion

In ideal situation, reinforcement concrete rebar corrosion was not happening but in actual situation the pore water may always be contaminated with different amounts of aggressive ions. This situation can be reproduced by the addition of 3.5% NaCl in the test solution (pore solution), with $NaNO_2$ (1, 2, 3, 4 and 5%) as inhibitor. **Figure 9** shows the polarization of rebar in pore solution +3.5% NaCl and various amounts (1, 2, 3, 4 and 5%) of $NaNO_2$. This figure shows the shift of potential of rebar in the positive direction in test solution (pore solution +3.5% NaCl) with addition of $NaNO_2$ (1–5%). This positive trend is due to rebuilding of rebar surface (rebuilding after local failure of oxide film by Cl^-) with the addition of $NaNO_2$. This is also indicative in **Figure 10**, where the corrosion current decreases with the addition of $NaNO_2$, and in **Figure 11** where the corrosion potential moves towards positive direction with the incremental addition of $NaNO_2$.

The addition of $NaNO_2$ in the experimental solution (PS + 3.5% NaCl) caused shift of both the anodic and cathodic curves towards positive direction as shown in **Figure 11**. Movement of polarization curves towards positive direction is indicative of the fact that the addition of $NaNO_2$ causes reduction in the rate of corrosion [1].

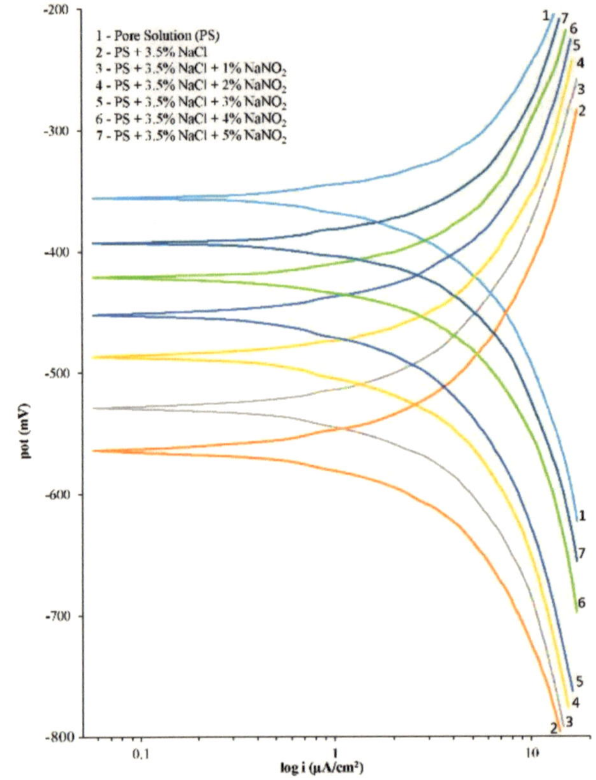

Figure 9.
Polarization of rebar in pore solution containing 3.5% NaCl and various $NaNO_2$ concentrations.

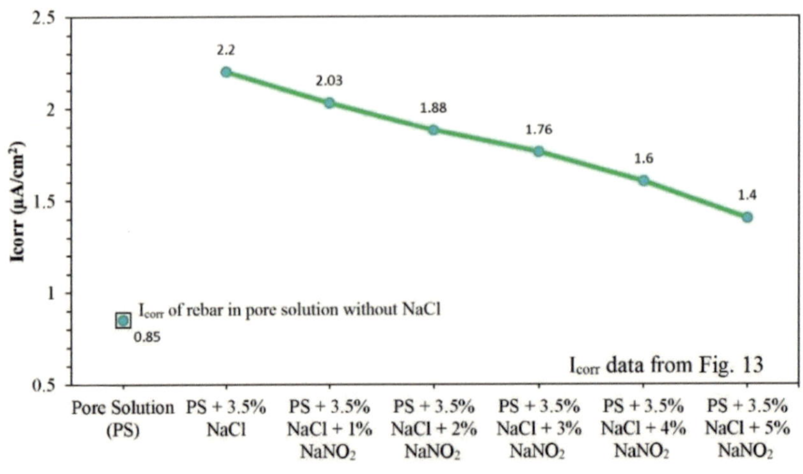

Figure 10.
I_{corr} for rebar (steel) in pore solution containing 3.5% NaCl and various $NaNO_2$.

3.2 Role of sodium nitrate ($NaNO_3$) inhibitor in rebar corrosion

Use of sodium nitrate ($NaNO_3$) as corrosion inhibitor has been in practice since recent past. In the present study, the rebar sample was exposed to pore solution containing 3.5% NaCl with further addition of 1–5% $NaNO_3$. The polarization diagram has been shown in **Figures 12** and **13**.

DOI: http://dx.doi.org/10.5772/intechopen.92374

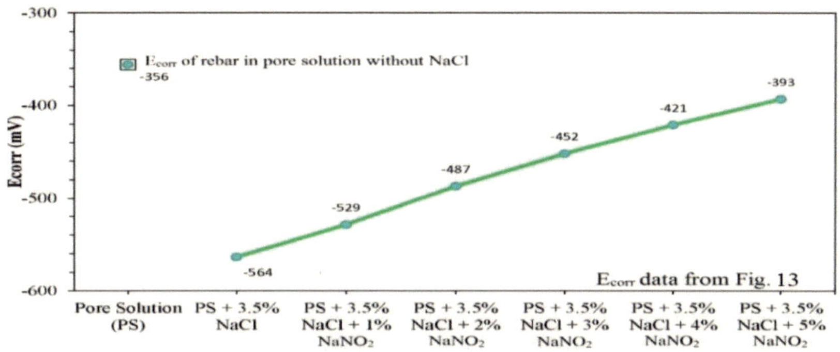

Figure 11.
E_{corr} for rebar (steel) in pore solution containing 3.5% NaCl and various $NaNO_2$ concentrations.

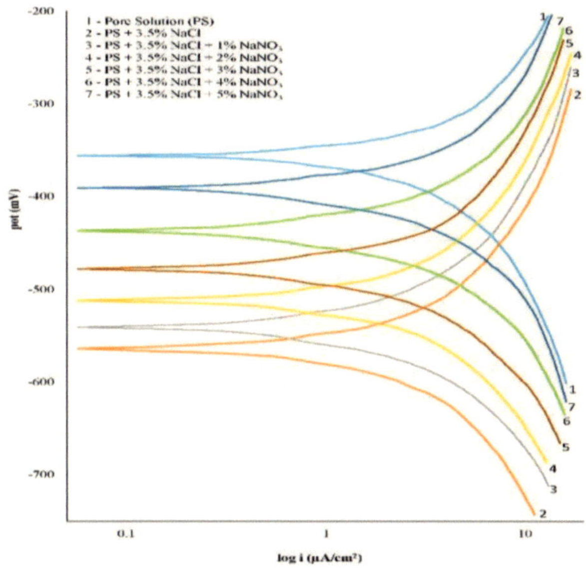

Figure 12.
Polarization of rebar in pore solution containing 3.5% NaCl and various $NaNO_3$ concentrations.

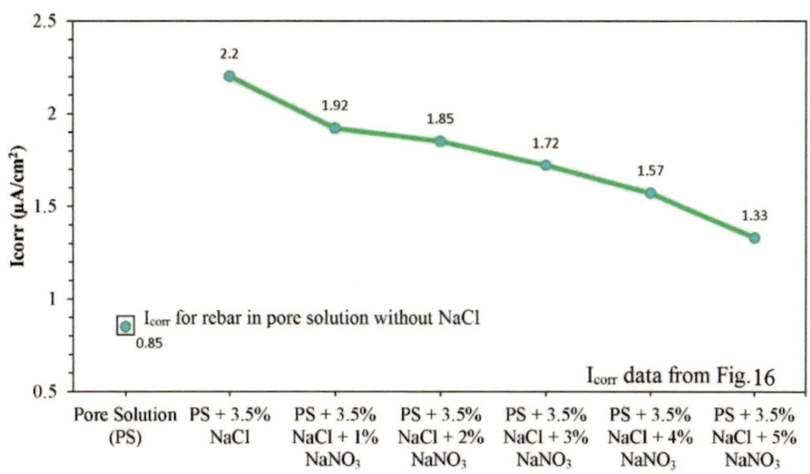

Figure 13.
I_{corr} for rebar in pore solution containing 3.5% NaCl and various $NaNO_3$ concentrations.

The I_{corr} and E_{corr} values of polarization are shown separately in **Figures 14** and **15**. The E_{corr} value with the addition of $NaNO_3$ increases in the positive direction and I_{corr} decreases continuously. A comparison of I_{corr} values obtained in $NaNO_2$ and $NaNO_3$ are given in **Table 1**, which indicates better performance of $NaNO_3$ (i.e., continuous decrease in I_{corr} values) in the ranges studied.

The I_{corr} values for rebar in pore solution containing various % of $NaNO_2$ and $NaNO_3$ as shown in **Table 1** indicate that $NaNO_3$ shows better inhibitive efficiency than $NaNO_2$ [1]. This can be attributed to the fact that three moles of ferrous ion is oxidized to ferric ion per mole of nitrate as shown in Eqs. (1) and (2).

$$2Fe\,(OH)_2\,(s) + NO_3^- + H_2O \rightarrow 2Fe\,(OH)_3\,(s) + NO_2^- \quad (1)$$

$$Fe\,(OH)_2\,(s) + NO_2^- + H_2O \rightarrow Fe\,(OH)_3\,(s) + NO + OH^- \quad (2)$$

3.3 Role of calcium nitrite inhibitor in rebar corrosion

Calcium nitrite is most widely used corrosion inhibitor for the protection of steel reinforcement corrosion in concrete. Here, similar conditions were used by addition of 3.5% NaCl in the test solution (pore solution), with Ca $(NaNO_2)_2$ (1, 2, 3, 4 and 5%) as inhibitor. **Figure 15** shows the polarization of rebar in pore solution containing 3.5% NaCl and various amounts of Ca $(NO_2)_2$.

From the above results, the I_{corr} and E_{corr} values have been calculated. The value of I_{corr} decreases continuously with the increase in the % of Ca $(NO_2)_2$ and the value of E_{corr} moves towards positive direction with the increase in Ca $(NO_2)_2$ content [1]. Both the above data are suggestive of continuous inhibition by increase in Ca $(NO_2)_2$ content because of the formation of passive oxide file on the metal surface. Actually in presence of chlorides, the nitrite 'competes' with both the chloride and hydroxide ion for the free Fe^{2+} ions. Over time, nitrite and/or an alkaline environment free of chlorides will reduce the number of flaws in the protective film and thus decrease the number of available sites from which chloride complexes may formed [2, 3].

3.4 Role of calcium nitrate inhibitor in rebar corrosion

Polarization diagram for rebar dipped in pore solution +3.5% NaCl + varying amounts (1–5%) of calcium nitrate Ca $(NO_3)_2$ is shown in **Figure 16**.

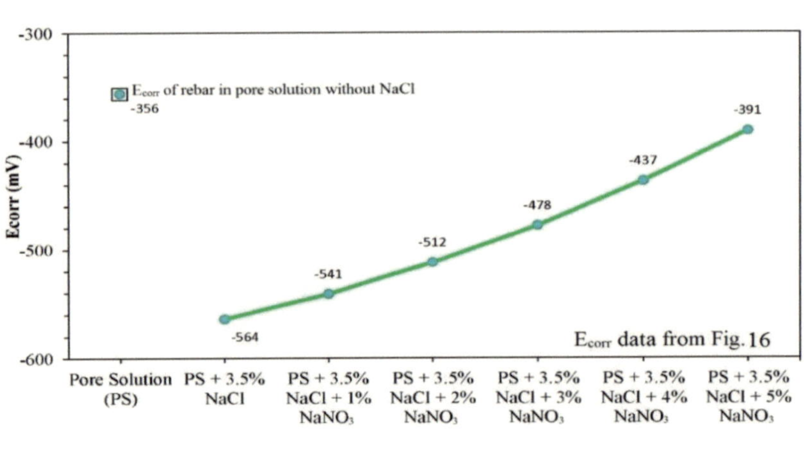

Figure 14.
E_{corr} for rebar in pore solution containing 3.5% NaCl and various $NaNO_3$ concentrations.

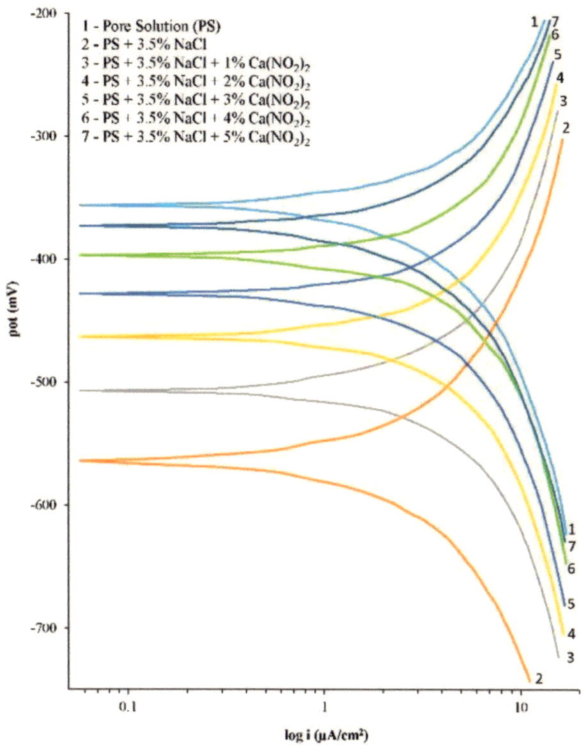

Figure 15.
Polarization of rebar in pore solution containing 3.5% NaCl and various Ca $(NO_2)_2$ concentrations.

Inhibitor	I_{corr} (μA/cm^2)				
	1%	2%	3%	4%	5%
$NaNO_2$	2.03	1.88	1.76	1.60	1.40
$NaNO_3$	1.92	1.85	1.72	1.57	1.33

Table 1.
I_{corr} values of rebar in PS containing 3.5% NaCl and various % of $NaNO_2$ and $NaNO_3$.

Figure 16 shows that the nature of protection of the calcium nitrate is comparable to that of calcium nitrite. A comparison of I_{corr} values of calcium nitrite and calcium nitrate as shown in **Table 2** indicates that calcium nitrate offers better inhibition within the range studied [1].

3.5 Role of sodium molybdate inhibitor in rebar corrosion

Sodium molybdate is a non-toxic [4], environment-friendly anodic corrosion inhibitor [5, 6]. It is highly effective in protecting steel reinforcement corrosion in concrete [7] and steel corrosion in saturated calcium hydroxide [8]. In actual situation the pore water in concrete may always be contaminated with different amounts of aggressive ions such as chloride. This situation can be represented by addition of 3.5% NaCl in the test solution (pore solution), with Na_2MoO_4 ($10^{-1.0}$, $10^{-1.5}$, $10^{-2.0}$, $10^{-2.5}$, $10^{-3.0}$, $10^{-3.5}$ and $10^{-4.0}$ M) as inhibitor. **Figure 17** shows the polarization of rebar in pore solution containing 3.5% NaCl and different concentrations of Na_2MoO_4.

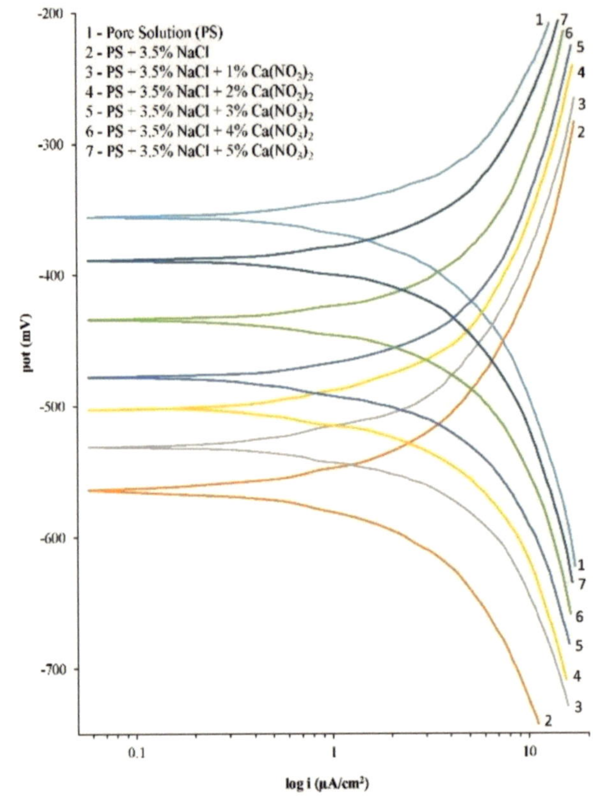

Figure 16.
Polarization of rebar in pore solution containing 3.5% NaCl and different Ca $(NO_3)_2$ concentrations.

Inhibitor	I_{corr} ($\mu A/cm^2$)				
	1%	2%	3%	4%	5%
$Ca(NO_2)_2$	1.93	1.81	1.71	1.58	1.31
$Ca(NO_3)_2$	1.88	1.79	1.66	1.51	1.24

Table 2.
I_{corr} values of rebar in PS containing 3.5% NaCl and different % of $CaNO_2$ and $CaNO_3$.

From the above results, the I_{corr} and E_{corr} values have been calculated. The I_{corr} decreases continuously with the increase in the molar concentration of Na_2MoO_4 and the value of E_{corr} moved towards positive direction with the increase in Na_2MoO_4 content [9]. Both the above data are suggestive of continuous inhibition by increase in Na_2MoO_4 content. This can be attributed to the formation of a hydrated mixed oxide film, which provides a barrier for anodic dissolution [10, 11]. In fact a non-protective ferrous (Fe^{2+})-molybdate complex is initially formed and is subsequently oxidized in the presence of oxygen [12]. The resulting ferric (Fe^{3+})-molybdate complex is insoluble and increases the stability of the Fe_2O_3 films that develop over the active corroding sites [13]. Further, the MoO_4^{2-} ions adsorb on the outermost part of the hydrated oxide layer (by ion exchange mechanism), thereby imparting a negative charge on the surface [14]. This leads to a barrier effect that impedes both the ingress of Cl^- ion to the underlying substrate and the transport of Fe^{2+} away from the surface [13].

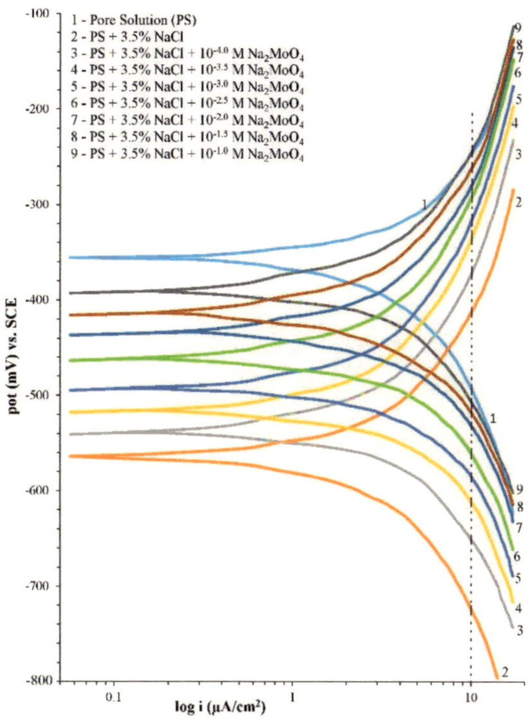

Figure 17.
Polarization of rebar (steel) in pore solution containing 3.5% NaCl and different concentrations of Na_2MoO_4.

3.6 Role of sodium tungstate inhibitor in rebar corrosion

Tungstates are very efficient corrosion inhibitor in more realistic environments such as neutral or alkaline solutions containing chlorides, sulphates and some other ions [15, 16]. Sodium tungstate has been extensively studied in the protection of iron as an environmentally-friendly anodic-type inorganic corrosion inhibitor [17–20] and is one of the effective corrosion inhibitors. The situation where pore water may always be contaminated with different amounts of aggressive ions such as chloride can be represented by addition of 3.5% NaCl in the test solution (pore solution), with Na_2WO_4 ($10^{-1.0}$, $10^{-1.5}$, $10^{-2.0}$, $10^{-2.5}$, $10^{-3.0}$, $10^{-3.5}$ and $10^{-4.0}$ M) as inhibitor. **Figure 18** shows the polarization of rebar in pore solution containing 3.5% NaCl and different concentrations of Na_2WO_4, which is almost similar to that in case of Na_2MoO_4 (**Figure 17**).

From the above results, the I_{corr} and E_{corr} values have been calculated. The value of I_{corr} decreases continuously with the increase in the molar concentration of Na_2WO_4 and the value of E_{corr} moves towards positive direction with the increase in Na_2WO_4 content [9]. Both the above data are suggestive of continuous inhibition by increase in Na_2WO_4 content. This may be attributed to the formation of a hydrated oxide film that acts a passive layer and provides a barrier for anodic dissolution [21, 22]. In fact the oxyanions (WO_4^{2-} ion) adsorb on the outermost part of the hydrated oxide layer, which leads to a barrier effect that inhibits both the ingress of Cl^- ion to the underlying substrate and the transport of Fe^{2+} away from the surface [23].

3.7 Comparative analysis of the inhibition by molybdate and tungstate

Inhibitors, molybdate and tungstate are oxygen-dependent anodic inhibitors [10–13, 24] in alkaline environment where these oxyanions remain stable and do

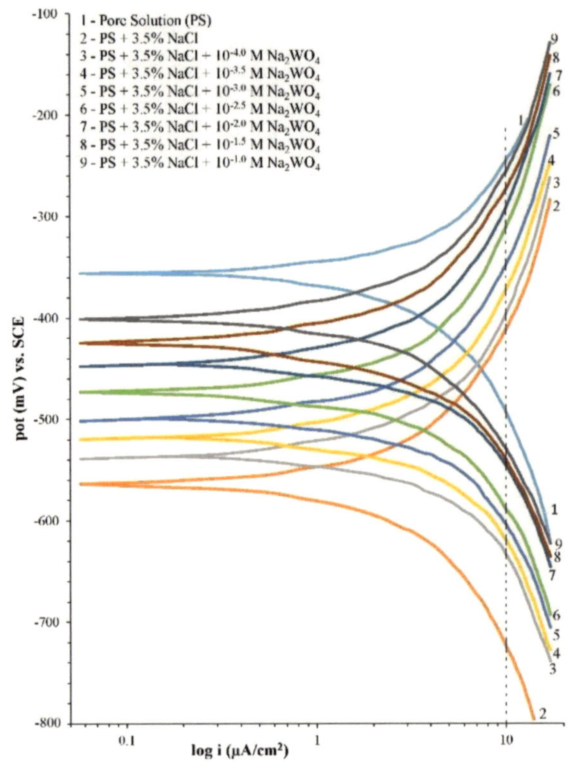

Figure 18.
Polarization of rebar (steel) in pore solution containing 3.5% NaCl and different concentrations of Na_2WO_4.

Inhibitor	I_{corr} (μA/cm^2)						
	$10^{-1.0}$ M	$10^{1.5}$ M	$10^{-2.0}$ M	$10^{2.5}$ M	$10^{3.0}$ M	$10^{3.5}$ M	$10^{-4.0}$ M
Na_2MoO_4	1.99	1.91	1.83	1.78	1.61	1.48	1.43
Na_2WO_4	1.98	1.92	1.82	1.77	1.60	1.51	1.41

Table 3.
I_{corr} values of rebar (steel) in pore solution containing 3.5% NaCl and different molar concentrations of Na_2MoO_4 and Na_2WO_4.

not promote spontaneous passivation in absence of dissolved oxygen [9]. The mechanism of inhibition involves the formation of a mixed oxide film, which acts as a barrier for anodic dissolution [13]. This has been reported that both molybdate and tungstate are comparable in their inhibitive efficiencies in alkaline environment [24] as obtained in the **Figures 17** and **18**.

However, the actual comparative performance of the sodium molybdate and sodium tungstate can be realized better from the I_{corr} values as given in **Table 3**. The I_{corr} values indicate that sodium molybdate and sodium tungstate are comparable in their inhibitive efficiencies; however, the inhibitive performance of sodium tungstate is slightly better [9].

4. Comparative analysis of the inhibition by nitrites (NO_2^-) and nitrates (NO_3)

Further to the discussion regarding inhibitive role of nitrite [Eqs. (1) and (2)], many other [25–27] theories have been put forward, some [28] even claimed that nitrite ions readily form complexing agents with Fe^{2+}, which are competitive with Cl^- complexation, thereby preventing formation of chlorocomplexes (these complexes appeared to be essential components of anodic corrosion processes [29]). Another proposal [27] explains sacrificial reduction of nitrite to nitrogen with simultaneous oxidation of ferrous to ferric ion. The formation of FeOOH (or similar product) has been indicated to decrease ion migration as ferrous/ferric chlorocomplexes and thereby stifle iron dissolution. It has been noted [25] that the reduction mechanism of NO_2^- inhibitor does not proceed further to nitric oxide (NO).

The inhibition process of NO_3^- may be summarized as: nitrate is reduced to nitrite [25] by ferrous ion in alkaline environment [pore solution].

$$2\,Fe(OH)_2\,(s) + NO_3^- + H_2O = 2Fe(OH)_3\,(s) + NO_2^- \qquad (3)$$

To explain the comparative performance of nitrite and nitrate, it me be summed up that one mole of NO_3^- inhibitor can offer protection equivalent to three moles of NO_2^- inhibitor.

The above strong inhibitive effect of NO_3^- has been noted in the present experiments with sodium and calcium salts, both in terms of potential and polarization studies.

5. Comparison of effectiveness of calcium and sodium inhibitors

Figure 19 below gives the assorted values of the I_{corr} vs. wt % data of sodium nitrite $NaNO_2$, sodium nitrate $NaNO_3$, calcium nitrite $Ca(NO_2)_2$ and calcium nitrate

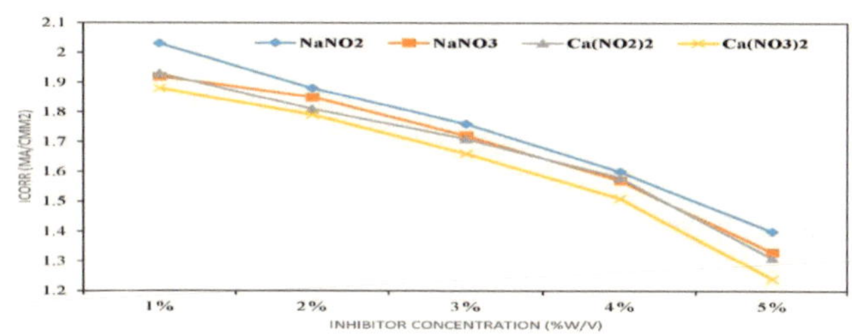

Figure 19.
Comparison of I_{corr} in terms of % m/v of inhibitors.

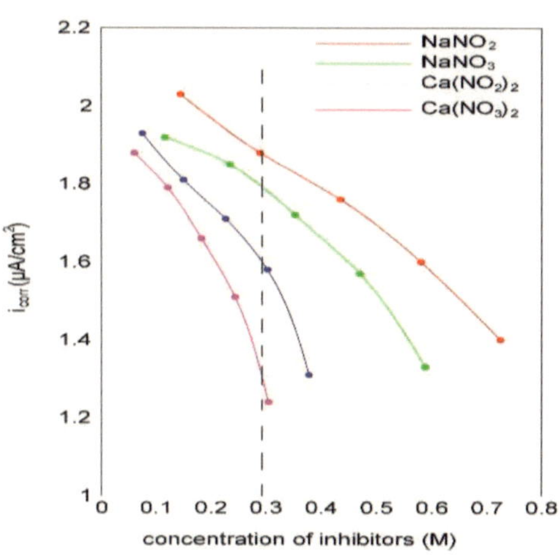

Figure 20.
Comparison of I_{corr} in terms of molar concentration of inhibitors.

$Ca(NO_3)_2$, which gives a comparative performance of sodium and calcium inhibitors. However, the actual comparative performance of the calcium and sodium inhibitors can be realized better if the molar concentrations are considered as in **Figure 20**, which indicates that for inhibition purpose, increasing the molar content, the calcium inhibitors become more effective (as marked in vertical dotted line).

6. Conclusion

The present investigation tried to explore types of inhibitors, nature of inhibitors, their mechanism and also to explore whether the simple polarization technique can be used to compare the efficiency of most prevalent inhibitors, for example, $NaNO_2$ and $NaNO_3$ as compared to the corresponding calcium salts and that of oxy-anions of group VI (molybdate and tungstate) for the inhibition of rebar in concrete. The findings indicate that performance wise (a) nitrate salts are more efficient than nitrite salts (sodium and calcium); (b) calcium salts are more effective; (c) further, it was observed that molar concentration wise calcium salts offer greater efficiency; and (d) molybdates and tungstates are almost comparable in their inhibitive efficiency in terms of I_{corr} values, however, tungstates show comparably better inhibitive efficiency in protecting steel reinforcement corrosion in concrete.

Acknowledgements

I would like to expresses my thanks to the Director, National Institute of Foundry and Forge Technology (NIIFT), Hatia, Ranchi, India, for their support and encouragement. Finally, I thanks the members of my family for their support and encouragement.

Author details

Anil Kumar
Department of Applied Sciences and Humanities, National Institute of Foundry and Forge Technology, Ranchi, India

*Address all correspondence to: akiitk_70@rediffmail.com

IntechOpen

© 2020 The Author(s). Licensee IntechOpen. This chapter is distributed under the terms of the Creative Commons Attribution License (http://creativecommons.org/licenses/by/3.0), which permits unrestricted use, distribution, and reproduction in any medium, provided the original work is properly cited.

References

[1] Ahamad A, Kumar A. Comparative behaviour of nitrite and nitrate for the protection of rebar corrosion. Journal of the Institution of Engineers (India): Series D. 2017;**98**(2):177-187

[2] El-Jazairi B, Berke NS. The use of calcium nitrite as a corrosion inhibiting admixture to steel reinforcement in concrete. In: Page CL, Treadaway KW, Bamforth PB, editors. Corrosion of Reinforcement in Concrete. London: Elsevier Science Publishers; 2002. pp. 571-585

[3] Berke NS, Weil TG. World-wide review of corrosion inhibitors in concrete. In: Malhotra VM, editor. Advances in Concrete Technology. 2nd ed. Ottawa: CANMET; 1992. pp. 891-914

[4] Vukasovich MS, Farr JPG. Molybdate in corrosion inhibitor –a review. Polyhedron. 1986;**5**(1/2):551-559

[5] Armour AW, Robitaille DR. Corrosion inhibition by sodium molybdate. Journal of Chemical Technology and Biotechnology. 1979;**29**: 619-628

[6] Weber TR, Stranick MA, Vukasovich MS. Molybdate corrosion inhibition in deaerated and low-oxygen waters. Corrosion. 1986;**42**:542-545

[7] Hope BB, Ip AKC. Effect of calcium nitrite and sodium molybdate on corrosion inhibition of steel in simulated concrete environment. Admixtures for Concrete—Improvement of Properties. Proceedings of the International Symposium Held by RILEM. New York, London: Chapman & Hall; 1990. ISBN: 0 412 37410 2 (Print edition)

[8] Tommaselli MAG, Mariano NA, Kuri SE. Effectiveness of corrosion inhibitors in saturated calcium hydroxide solutions acidified by acid rain components. Construction and Building Materials. 2009;**23**:328-333

[9] Ahmad A, Kumar R, Kumar A. Effect of sodium molybdate and sodium tungstate in concrete rebar corrosion. Anti-Corrosion Methods and Materials. 2019;**63**(3):253-263

[10] Pryor MJ, Cohen M. The inhibition of the corrosion of iron by some anodic inhibitors. Journal of the Electrochemical Society. 1953;**100**(5):203-215

[11] Vukasovich MS, Robitaille DR. Corrosion inhibition by sodium molybdate. Journal of the Less Common Metals. 1977;**54**(2):437-448

[12] Lizlovs EA. Molybdates as corrosion inhibitors in presence of chlorides. Corrosion NACE. 1976;**32**(7):263-266

[13] Kodama T, Ambrose JR. Effect of molybdate ion on the repassovation kinetics of iron in solutions containing chloride ions. Corrosion. 1977;**33**:155-161

[14] Al-Saffar AH, Ashworth V, Bairamov AKO, Chivers DJ, Grant WA, Procter RPM. The effect of molybdenum ion implantation on the general and pitting corrosion behaviour of pure aluminium and a high strength aluminium alloy. Corrosion Science. 1980;**20**:127-144

[15] Mammoloti L, Hansson CM, Hope BB. Corrosion admixtures in concrete part II: Effect on chloride threshold values for corrosion of steel in synthetic pore solutions. Cement and Concrete Research. 1999;**29**(10):1583-1589

[16] Cisse MB, Zergar B, El Kalai F, EbnTouhami M, Sfaira M, Taleb M, et al. Two dipodal pyridine-pyrazol derivatives as efficient inhibitors of mild steel corrosion inHCl solution—part I: Electrochemical study. Surface Review and Letters. 2011;**18**(6):303-313

[17] Shibli SMA, Saji VS. Co-inhibition characteristics of sodium tungstate with potassium iodate on mild steel corrosion. Corrosion Science. 2005;**47**: 2213-2224

[18] Boucherit MN, Amzert SA, Arbaoui F, Hanini S, Hammache A. Pitting corrosion in presence of inhibitors and oxidants. Anti-Corrosion Methods and Materials. 2008;**55**:115-122

[19] Rao BVA, Rao MV, Rao SS, Sreedhar B. Tungstate as a synergist to phosphonate-based formulation for corrosion control of carbon steel in nearly neutral aqueous environment. Journal of Chemical Sciences. 2010;**122**(4):639-649

[20] Alentejano CR, Aoki IV. Localized corrosion inhibition of 304 stainless steel in pure water by oxyanions tungstate and molybdate. Electrochimica Acta. 2004;**49**:2779-2785

[21] Celeste RA, Vieira VA. Localized corrosion inhibition of stainless steel in pure water by oxyanions tungstate and molybdate. Electrochimica Acta. 2004;**49**:2779-2785

[22] Song Q, Qui XQ. Study on the complex of sodium tungstate and urotropine as inhibitors against stainless steel corrosion in the NaCl solution. Materials Science. 2011;**689**:450-454

[23] Deepak VK, Jagadeesh B. Effect of sodium tungstate as a green corrosion inhibitor on the passivation behaviour of mild steel sheet in aggressive media. International Journal of Applied Sciences and Biotechnology. 2016;**4**(2):183-190

[24] Robertson WD. Molybdate and tungstate as corrosion inhibitors and the mechanism of inhibition. Journal of the Electrochemical Society. 1951;**98**(3):94-100

[25] Rosenberg AM, Gaidis JM. The mechanism of nitrite inhibition of chloride attack on reinforcing steel in alkaline aqueous environments. Materials Performance. 1979;**18**(11):45-48

[26] Gaidis JM, Rosenberg AM. The inhibition of chloride-induced corrosion in reinforced concrete by calcium nitrite. Cement, Concrete and Aggregates. 1987;**9**(1):30-33

[27] El-Jazairi B, Berke NS. The use of calcium nitrite as a corrosion inhbiting admixture to steel reinforcement. Corrosion of Reinforcement in concrete. In: Page CL, editor. Society of Chemical Industry; London. 1990. pp. 571-585

[28] Sagoe-Crentsil KK, Yilmaz VT, Glasser FP. Properties of inorganic corrosion inhibition admixtures in steel-containing OPC mortars. Part 1: Chemical properties. Advances in Cement Research. 1992;**4**:91-96

[29] Sagoe-Crentsil KK, Glasser FP. Green rust, iron solubility and the role of chloride in the corrosion of steel at high pH. Cement and Concrete Research. 1993;**23**:785-791

Section 2

Coatings

Chapter 4

Organic-Inorganic Hybrid Coatings for Active and Passive Corrosion Protection

Andressa Trentin, Samarah V. Harb, Thiago A.C. de Souza, Mayara C. Uvida and Peter Hammer

Abstract

Organic-inorganic coatings based on poly(methyl methacrylate) (PMMA)-silica and PMMA-cerium oxide hybrids provide effective and active corrosion protection of metallic surfaces. For both hybrid materials, the covalent conjugation of inorganic silica or ceria nanodomains with the PMMA matrix, provided by molecular coupling agents, leads to homogenous and highly cross-linked nanocomposites, which act in the form of coatings as an efficient diffusion barrier. The addition of lithium salts (500–2000 ppm) into PMMA-silica hybrid and optimized ceria fraction in PMMA-cerium oxide coatings results in active corrosion inhibition by the self-healing effect. Results of electrochemical assays of aluminum- and steel-coated samples, performed in a 3.5% NaCl solution, show an excellent corrosion resistance (impedance modulus up to 100 GΩ cm^2) and durability (up to 350 days) of the 10-μm-thick passive barrier layer. Time-of-flight secondary ion mass and X-ray photoelectron spectroscopies evidenced the self-healing ability of coatings induced by lithium/cerium ion leaching toward corrosion spots or artificial scratches, which are restored by a protective layer of precipitated phases. Results presented in this book chapter evidence the active role of lithium and cerium species in improving the hybrid structure and providing through self-healing a significantly extended service life of metallic components.

Keywords: anticorrosion coating, organic-inorganic hybrid, passive and active protection, corrosion inhibitors, self-healing, AA7075, carbon steel

1. Introduction

The development of efficient production techniques of metallic alloys in the late twentieth century formed the basis for the boom of construction, transportation, energy, packaging, and electronic industries [1, 2]. Although extensively used, unprotected alloys are subject to a constant oxidation process, either in contact with humid atmosphere, aqueous, or soil systems, turning them into natural ore [1, 3]. One common approach to prevent or at least delay metallic corrosion is the application of organic coatings as a physical diffusion barrier. A barrier coating in the form of a dense insulating layer acts as a quasi-ideal capacitor, which inhibits the contact of corrosive species, such as the electrolyte, with the substrate. Barrier coatings based on acrylate, epoxy, and polyurethane are extensively employed in automotive, aviation, and marine industries but also in electronics, decoration, food, and beverage items. However, if these coating systems are not designed to withstand aggressive environments, they fail, causing corrosion of the underlying alloy. Therefore, conventional high-efficiency coatings are prepared in the form of multilayers combining a chromate conversion layer (0.1–0.2 μm) with a primer (commonly an epoxy layer of about 15–25 μm loaded with corrosion inhibitors) and an organic topcoat, mostly in the form of a 50–100-μm-thick polyurethane, which provide barrier, decorative, hydrophobic, and UV-resistant characteristics [4, 5].

Chromates, widely employed as conversion layer in the aerospace industry, provide effective protection because the excess of non-reduced ions produces an extremely corrosion-resistant film composed of mixed Cr and Al oxides [6]. However, hexavalent chromium causes occupational health problems due to carcinogenic and mutagenic effects [7], which led to an intense search for high-performance chromium-free coatings able to protect passively and actively metal surfaces in harsh environments. To achieve this goal, the alternative material must provide (i) a dense passive barrier with a very low permeation rate and after failure (ii) actively inhibit corrosive processes with similar efficiency as the self-healing ability of chromate anions. Considering this challenge, considerable efforts have been spent by the scientific community to develop alternatives to chromium conversion coatings, and some encouraging results have been already achieved.

One promising class of coating systems, developed in the last decade, which fulfill the cited criteria, is organic-inorganic nanocomposites based on conventional acrylic, epoxy, and polyurethane materials combined with ceramic nanofillers, such as silica, ceria, zirconia, etc. These hybrid materials have demonstrated excellent barrier property, providing long-term protection for steel and aluminum alloys [3, 8–10]. The superiority of hybrid systems compared to purely polymeric phases comes from a tailored nanostructure achieved by proper amounts of inorganic nanofillers within the organic matrix and the careful tuning of synthesis conditions, resulting in a dense and homogeneous nanocomposite that acts as an efficient diffusion barrier, limiting the water uptake and diffusion of ionic species to a very low rate [11]. The inorganic nodes have the important role to densify the structure by anchoring covalently the polymeric chain segments through a cross-linking agent and to improve the adhesion at the coating/metal interface through covalent bonding [12–14]. The function of the polymeric phase is to provide minimum internal stress and porosity of the hybrid network, hermetically sealing the structure [3, 11, 12]. Hence, the key factor for the stability of the material under adverse conditions is the covalent conjugation between the two phases through the coupling molecule that has functional similarity to both parts. The presence of this organic-inorganic interphase overcomes the limitations of organic systems such as lack of thermal and mechanical stability, poor adhesion, and presence of voids due to material swelling [15–18].

Nevertheless, even the best barrier fails after some time due to the permeation of water and aggressive species under weathering or mechanical damage. Smart coatings are an emerging technology to bypass the limitations of passive systems, providing a repair response mechanism after failure [5]. Several approaches have been developed in the search for chromium-free organic coatings using organic and inorganic additives for extrinsic self-healing strategies. These types of coatings are able to respond after local damage through reactions triggered by variations in pH, temperature, presence of water, mechanical damage, and UV [3, 4, 19].

In practice, however, some studied coating systems, especially those based on inhibitor-filled micro- and nano-containers, imply increasing costs, which are hardly accepted by the industry. Some recent studies reporting on incorporation of simple additives, such as PANI [20], TiO_2 [21], ZnO [22], lithium [23, 24], ZrO_2 [9, 25], cerium/ceria [26–29], poly(2-butylaniline) (P2BA) [30], polydopamine (PDA) [31], 2-mercaptobenzimidazole (MBI) [32], and tannins [33], among others, have shown promising results toward the development of chromium-free active coatings. This chapter comprises an overview of efficient passive organic-inorganic coatings and their modified form for active protection of metal surfaces. Emphasis is given on inorganic additives that meet the current demand for non-toxic low-cost substances, such as cerium nanoparticles and lithium ions in hybrid matrices.

2. Organic-inorganic hybrid coatings

2.1 Passive protection

Organic-inorganic coatings are usually prepared by combining the polymerization of the organic phase using an anionic, cationic, thermal, or photo initiator with the sol-gel route of hydrolysis and condensation of an inorganic compound in the form of silicon, zirconium, aluminum, and titanium alkoxides [34]. In the next step, the hybrid sol is applied on the metal surface by spray, spin, or dip coating methods, and then the obtained film is dried and cured prior to the structural, thermal, mechanical, and electrochemical analysis.

Epoxy [35–38] and polyurethanes-based [21, 39, 40] hybrid coatings present excellent protection against metallic corrosion; however recent research has shown that equal or even better results in terms of corrosion resistance and durability can be achieved with much thinner layers, based on acrylic or epoxy hybrids. Especially for poly(methyl methacrylate) (PMMA)-silica coatings, remarkable results have been reported when applied on carbon steel and aluminum alloys. For instance, using electrochemical impedance spectroscopy (EIS) assays, Hammer et al. reported that PMMA-silica coatings with thicknesses between 1.5 and 3 μm are able to withstand adverse conditions without failure for 18 days immersed in 3.5% NaCl, while maintaining the low-frequency impedance modulus ($|Z_{lf}|$) in the GΩ cm^2 range, a value about five orders of magnitude higher than that of bare steel [41]. Analyzing the structural properties of this material, dos Santos et al. showed that optimizing the inorganic solvent proportion (ethanol/H_2O ratio) results in higher connectivity of the silica phase, leading to an improvement of the corrosion protection ($|Z_{lf}| > 1$ GΩ cm^2) and durability up to 6 months in 3.5% NaCl and more than 3 months in saline/acid solution [12].

Excellent passive corrosion protection of carbon steel was recently reported for some micrometer-thick PMMA-silica coatings in a study where the ratio of thermal initiator (benzoyl peroxide (BPO)) to monomer (methyl methacrylate (MMA)) was varied in the range of 0.01–0.1 (B001, B005, B010 samples) [11]. The authors found that this parameter strongly influences the hybrid structure in terms of polymerization efficiency, leading to an improved anticorrosion performance. From Raman spectra (**Figure 1a**), it can be observed that as the BPO fraction increased, the C=C band at 1640 cm^{-1} disappeared, indicating that the larger number of radicals provided by BPO promotes a more efficient polymerization of MMA. This result is supported by thermogravimetry, carried out under N_2 atmosphere to analyze the decomposition stages of the hybrid structure (**Figure 1b**). The first derivative of the thermogravimetric (TG) curves (**Figure 1c**) shows that the PMMA phase has three major degradation events, involving the rupture of head-to-head segments (stacking defects) at about 240°C (T_1), breaking of unsaturated chain ends at ~300°C (T_2), and a random breaking of head-tail segments at ~400°C (T_3). The differential thermogravimetric (DTG) curves clearly show that the stacking defects (T_1 event) are strongly suppressed for higher BPO/MMA ratios, evidencing improved polymerization. The residual mass of about 20 wt% found at 800°C corresponds to the nominal fraction of the silica phase in the hybrid.

In this study it was shown that the combination of adequate proportions between reactive groups of organic monomer (MMA), silica precursor (tetraethoxysilane (TEOS)), and coupling agent (3-(trimethoxysilyl)propyl methacrylate (MPTS)) leads to the formation of a homogeneous and defect-free structure

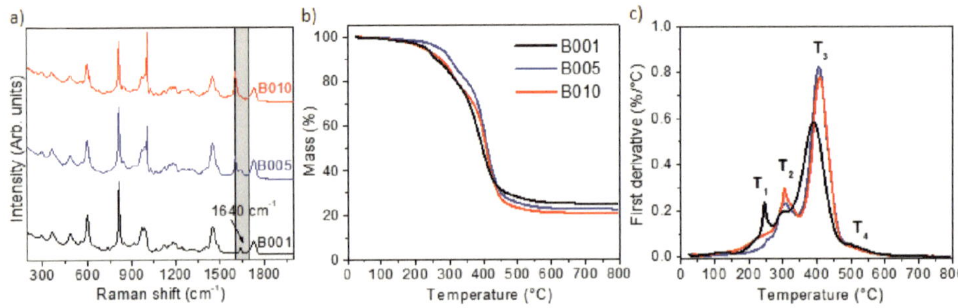

Figure 1.
(a) Raman spectra, (b) TG curves and (c) DTG curves of PMMA-silica hybrids prepared using BPO/MMA ratios of 0.01, 0.05, and 0.10 (reproduced with permission from Elsevier [11]).

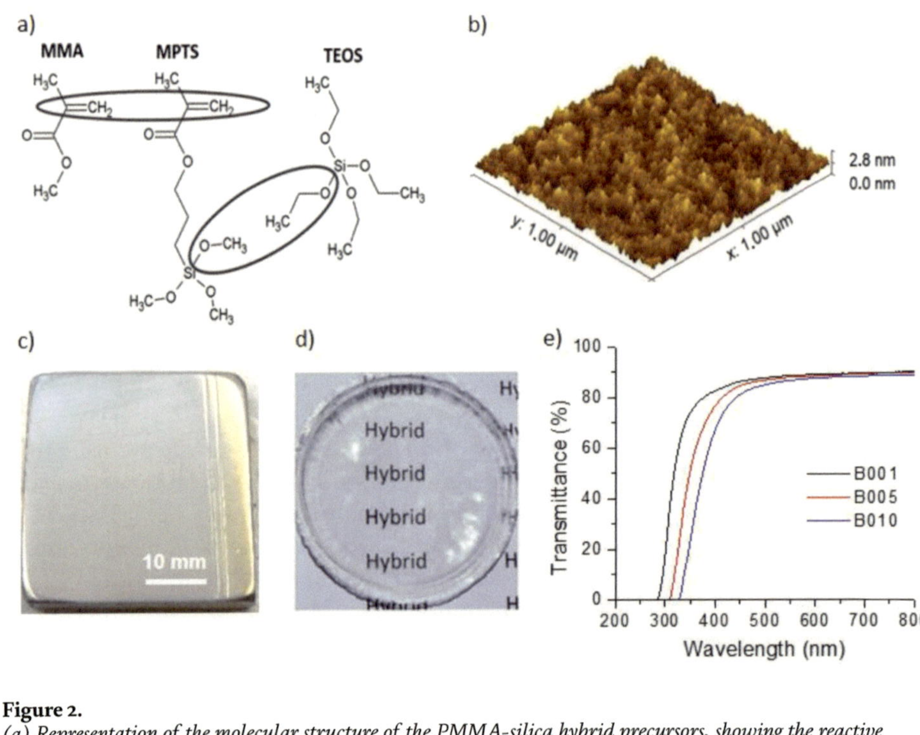

Figure 2.
(a) Representation of the molecular structure of the PMMA-silica hybrid precursors, showing the reactive groups of MMA and silicon alkoxides; (b) 3D AFM topography image of the B001 coating deposited on carbon steel; (c) representative images of the coated carbon steel; (d) unsupported hybrid; (e) UV–vis transmittance spectra of the unsupported hybrids (reproduced with permission from Elsevier [11]).

(**Figure 2a**). According to atomic force microscopy (AFM), optical, and UV–vis results (**Figure 2b–e**), the prepared coatings are transparent (80–90% transmittance) and very smooth (RMS roughness <1 nm), expected features for effective barrier coatings. For increasing amounts of the thermal initiator, the red shift of the absorption edge (**Figure 2e**) may be related to increased absorption of BPO in the UVA spectral range. Furthermore, adhesion pull-off tests evidenced a strong adherence of the coatings to the steel substrate, reaching values up to 26.3 MPa for the B001 sample, while 14.8 MPa, 8.9 MPa, and 6.7 MPa were determined for the B005, B010 and pure PMMA, respectively.

To evaluate the anticorrosion performance of coated and uncoated steel, EIS measurements were performed in duplicate after immersion of the coatings in neutral 3.5% NaCl solution at regular intervals until a significant drop in the impedance modulus was observed, representing the lifespan of the coating. After 3 h of immersion, all hybrid coatings showed a quasi-ideal capacitive behavior over almost the entire frequency range, contrasting with the poor performance of pure PMMA and bare steel (**Figure 3a**).

The comparison with pure PMMA coating shows clearly the crucial role of the silica phase in increasing corrosion resistance, after 1 day of immersion, from a low-frequency impedance modulus of 200 kΩ cm^2 (pure PMMA) to more than 5.0 GΩ cm^2 (PMMA-silica coatings). Moreover, the BPO increase led to a considerable lifespan extension for the B010 coating (9.1 μm thick) reaching 583 days with a nearly unchanged electrochemical response (**Figure 3b**), compared to the lifetime of 40–50 days, observed for the B005 and B001 coatings (5.0 and 2.8 μm thick, respectively), and only 1 day of the PMMA film. This anticorrosion

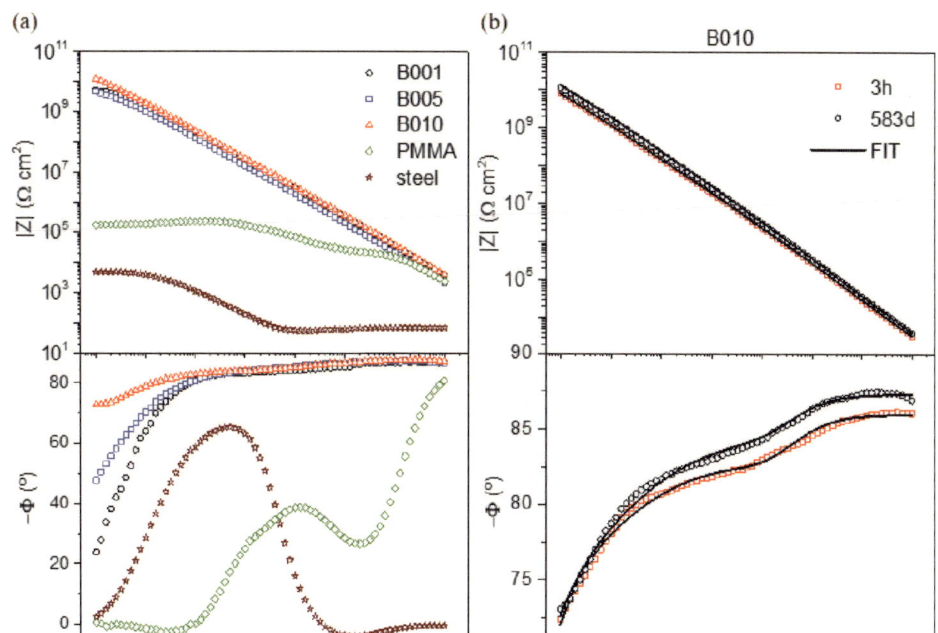

Figure 3.
Bode plots recorded (a) after 3 h immersion in 3.5% NaCl, for PMMA-silica coatings on carbon steel prepared at different BPO/MMA ratios, pure PMMA (BPO/MMA = 0.01) and bare steel and (b) for different immersion times of the B010 sample. The symbols represent the experimental data, and the solid lines are fits obtained using the electrical equivalent circuit (EEC) model (reproduced with permission from Elsevier [11]).

Coating	Substrate	Thickness (μm)	$\lvert Z_{lf} \rvert$ (GΩ cm^2) lifetime (days) solution	Ref.
PMMA-MPTS-TEOS	A1010 carbon steel	2.3–9.7	~5, 583, 3.5% NaCl	[11]
PMMA-MPTS-TEOS	AA2024	~3	~50, >560, 3.5% NaCl	[3]
PMMA-MPTS-TEOS	A1010 carbon steel	1.5–2	~5, 196, 3.5% NaCl	[12]
Acrylic resin-SiO_2	Mild steel	75	~10, 90, 3.5% NaCl	[43]
PMMA-MPTS-TEOS-CNTs or GO	A1010 carbon steel	3–6	~3, 211, 3.5% NaCl	[42]
Epoxy-zinc phosphate-iron oxide	Cold rolled low carbon steel	55–140	~450/60, 220/405, 3% NaCl	[8]
Epoxy-PANI	Mild steel	20	10–100, 120, 3.5% NaCl	[20]
Epoxy-GO-P2BA	Q235 carbon steel	20	10, 80, 3.5% NaCl	[30]
PU-PANI	Mild Steel	~60	10, 58, 3.5% NaCl	[40]
PU-ZrO_2-SiO_2	Carbon steel	40–55	~100, 226, 3.5% NaCl	[9]
PU-MMT	Carbon steel	~40	~10, 225, 3.5% NaCl	[39]
PU-PS-PLA-MBT	AA2024	152	~0.1, 83, 3.5% NaCl	[44]
PU-TEOS-TIP-ZRP	AA2024	203	~10, 100, 3.5% NaCl	[45]

PMMA, poly(methyl methacrylate); MPTS, 3-(trimethoxysilyl)propyl methacrylate; TEOS, tetraethoxysilane; CNTs, carbon nanotubes; GO, graphene oxide; PANI, polyaniline; MMT, montmorillonite; P2BA, poly (2-butylaniline); PU, polyurethane; PS, polysiloxane; PLA, polylactic acid; MBT, 2-mercaptobenzothiazole; TIP, titanium(IV) isopropoxide; ZRP, zirconium(IV) propoxide.

Table 1.
Reported high-performance passive hybrid coatings: composition, substrate, thickness, impedance modulus at low frequency $\lvert Z_{lf} \rvert$, lifetime, and solution.

performance, associated with an improved structure (**Figure 1**), is comparable to that of the best performing anticorrosive coatings reported so far [8, 9, 12, 20, 21, 39, 42]. Results obtained for high-performance passive barriers based on organic-inorganic coatings reported by several research groups are summarized in **Table 1**.

Additional information on the barrier property of the B010 sample was obtained by a cross-sectional analysis using scanning electron microscopy coupled with energy dispersive X-ray analysis (SEM/EDX) before and after 583 days of immersion. **Figure 4a** shows an essentially unaffected morphology, free of pores and defects, after 583 days of immersion. Further evidence comes from the EDS profiles recorded along normal axis of the coating, shown in **Figure 4b**. The spectrum of the immersed sample shows no chlorine signal, which would indicate the presence of Cl^- ions in the bulk, thus confirming the excellent barrier properties of this hybrid nanocomposite.

From these results, it can be concluded that the quantity of the thermal initiator plays a crucial role in terms of the connectivity of the organic phase, resulting in a remarkable improvement of the barrier property for the B010 coating (BPO/MMA = 0.1). The increase of the impedance modulus of more than five orders of magnitude and the long durability of this coatings, compared to the pure PMMA film, highlights the crucial role of the silica phase and the importance of an efficient polymerization. Considering the excellent performance achieved by PMMA-silica coatings without any additives, this material can be considered as a promising alternative for conventional primer systems for the protection of steel surfaces.

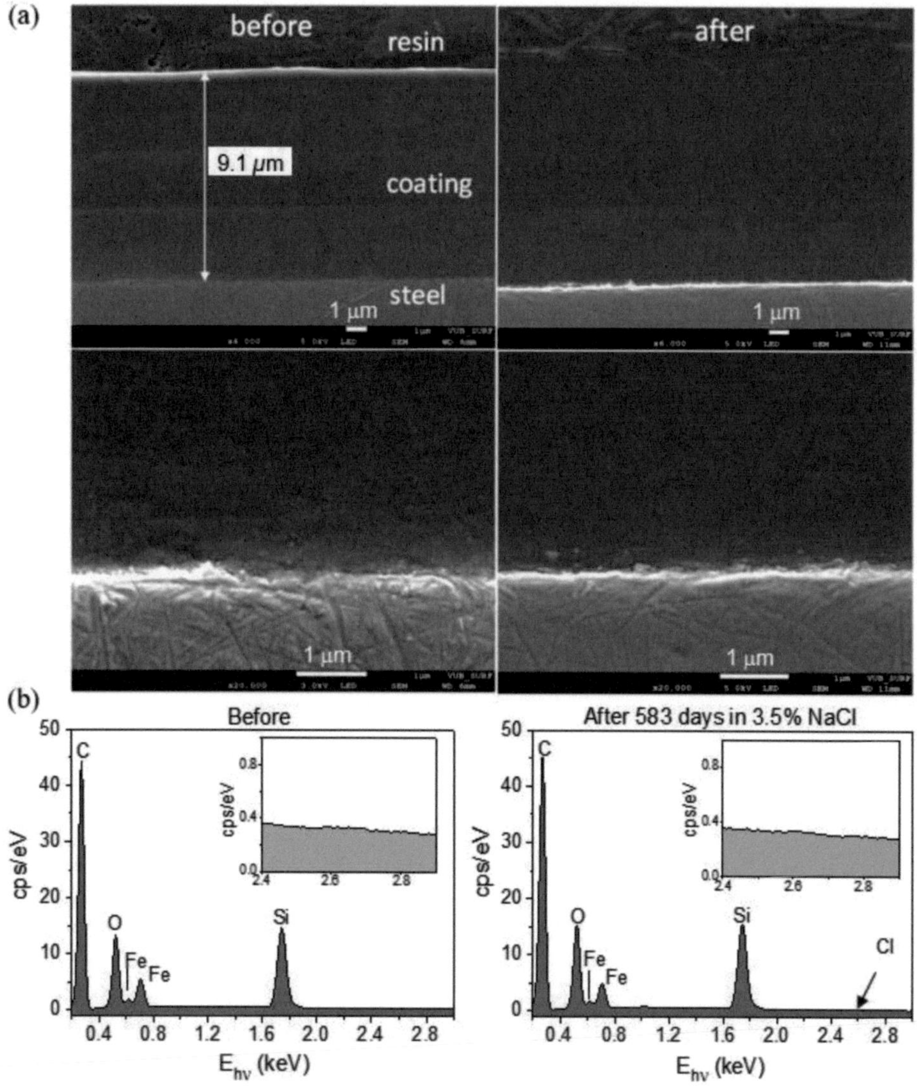

Figure 4.
(a) Cross-sectional SEM images of the B010 coating, before (left) and after (right) 583 days of immersion in 3.5% NaCl solution, including details of the coating/steel interface. The layers visible in the images are due to the three dips applied by dip coating; (b) integrated EDX cross-sectional spectra, before and after 583 days of immersion (reproduced with permission of Elsevier [11]).

2.2 Active protection

2.2.1 Corrosion inhibition by lithium ions

The second requirement for an effective chromium-free coating relies on active anticorrosion protection. Among corrosion inhibitors, it was demonstrated that the incorporation of lithium salts improves corrosion protection of epoxy coatings. As possible mechanisms, Visser et al. have proposed that the release of lithium ions from the epoxy matrix forms in the defective zone a Li/Al layer (hydrotalcite class) by a conversion process [46]. In recent work, lithium carbonate (Li_2CO_3) was for the first time added (500–2000 ppm) into the PMMA-silica system, and the results revealed a dual beneficial effect of lithium on the structure and self-healing ability of the coatings on the AA7075 aluminum alloy [24].

The structural analysis of the coatings showed that the changes of the small-angle X-ray scattering (SAXS) profiles imply significant modifications in the nanostructure of the silica phase for higher lithium loadings (**Figure 5a**). The profiles of Li0 (0 ppm) and Li05 (500 ppm) samples present a correlation peak, indicative of a concentrated set of silica domains with an average spacing of 3–4 nm (**Figure 5b**). However, for further addition of lithium (Li1 and Li2), the correlation peak vanishes due to the formation of a more diluted set of larger silica clusters. For these profiles, the average size of the silica nanoparticles (R_g) can be calculated according to the Guinier-Porod model [47, 48], revealing for the Li1 sample (1000 ppm) a gyration radius of about 1 nm and close to 3 nm for Li2 (2000 ppm).

Thermogravimetric measurements were used to access information on the polymerization efficacy of PMMA and the thermal stability of the material. The results reveal that the samples exhibited a thermal stability up to ~240°C (onset degradation temperature at 5% mass loss, T_s (**Figure 5c**)), while degradation events of the TG derivative curve (**Figure 5d**) showed that the presence of lithium suppresses head-to-head stacking defects (T_1) and unsaturated PMMA chain ends (T_2). These findings indicate a more efficient polymerization with higher Li content, which is an essential feature for an effective diffusion barrier.

The films with thickness between 4 μm and 6 μm present excellent adhesion to AA7075 substrate reaching values up to 28 MPa for coatings with higher Li loading [24]. The improved adhesion for lithium-rich coatings might be related to

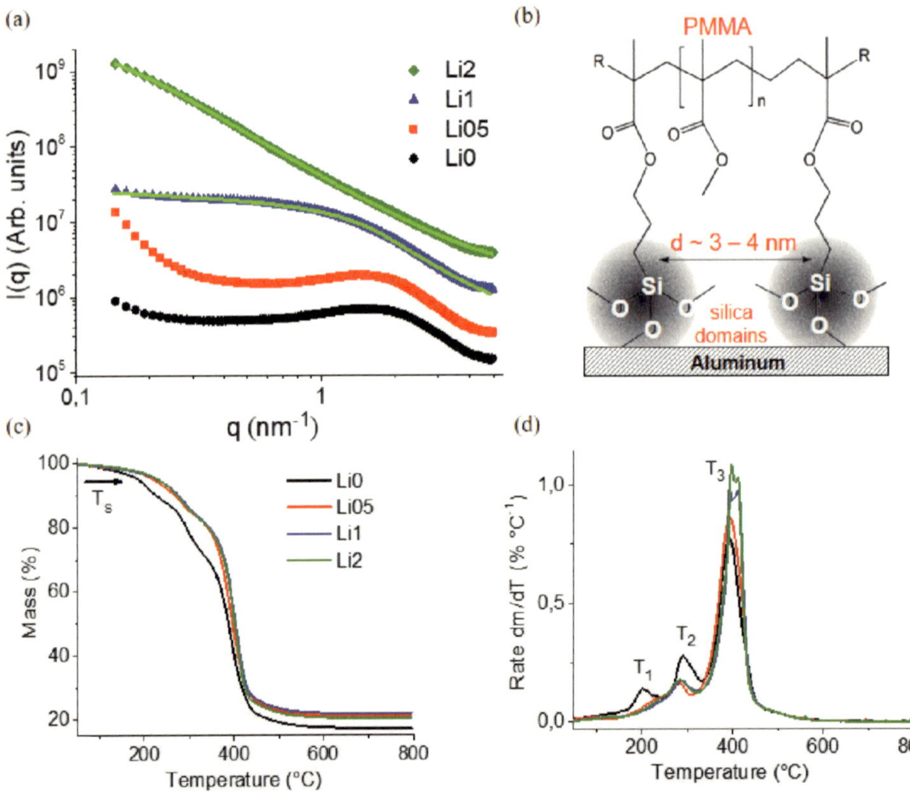

Figure 5.
(a) SAXS intensity profiles of the hybrids fitted according to the Guinier-Porod model (green lines); (b) structural representation of Li0 and Li05 PMMA-silica hybrids; (c) thermogravimetric curves and (d) differential thermogravimetric curves of PMMA-silica hybrids prepared with Li_2CO_3 concentrations of 0, 500, 1000, and 2000 ppm (reprinted (adapted) with permission from [24]. Copyright (2020) American Chemical Society).

aluminum enrichment near the coating/substrate interface, suggesting a diffusion of Al from the alloy into the coating during the thermal treatment [24].

Besides the beneficial effects of lithium on the structural properties, Li-modified PMMA-silica hybrids yielded coatings with longer durability under immersion in 3.5% NaCl solution. This is evident from the time dependence of the open circuit potential (E_{OCP}) and the low-frequency impedance modulus ($|Z_{lf}|$), recorded by EIS during 310 days of immersion (**Figure 6**). For lithium-containing coatings, the time dependence of $|Z_{lf}|$ (**Figure 6b**) revealed a very interesting effect of coating regeneration, attributed to the chemical activity of the lithium ions. The data show several important features of the lithium activity for different Li loadings: (i) for all samples the time dependence shows a clear correlation between $|Z_{lf}|$ and E_{OCP} (**Figure 6a**); (ii) after local failure of the coating, an impedance drop occurred, followed by a gradual recovery of about two orders of magnitude within 40 days (Li05), 28 days (Li1), and 20 days (Li2), indicating a faster regeneration process with increasing Li content; and (iii) the higher lithium concentration of Li2 coating delayed the appearance of localized corrosion compared to Li1 and Li05 samples. Furthermore, although the initial value of $|Z_{lf}|$ was smaller for Li1 and Li2 coatings, after 300 days of immersion, they showed a higher impedance value, an effect that can be related to their improved structural characteristics, as discussed before.

Representatively, the time evolution of the Bode plots obtained by EIS for the Li1 coating is displayed in **Figure 7**, together with the fitted curves (3 h, 99 days, and 126 days) using electrical equivalent circuits (EEC), shown in **Figure 7a**. Initially, Li1 presents an impedance modulus up to six orders of magnitude higher than bare aluminum alloy. As a consequence of the coating permeation by the electrolyte, the formation of conductive percolation paths leads after 99 days to the appearance of localized corrosion spots (pits), causing a decay of the impedance modulus at medium and low frequencies, indicative for failure of the coating and the beginning corrosion process at the coating/metal interface [11]. Nevertheless, after 126 days of immersion, the coating was spontaneously restored, increasing $|Z_{lf}|$ by two orders of magnitude. Next, the appearance of a second pit caused a new decay after 154 days, and thereafter the impedance modulus has recovered again after 183 days, remaining stable until 311 days of immersion.

The self-healing process of the first recovery event was analyzed by fitting the EIS data using EEC containing a electrolyte resistance in series with two time constants,

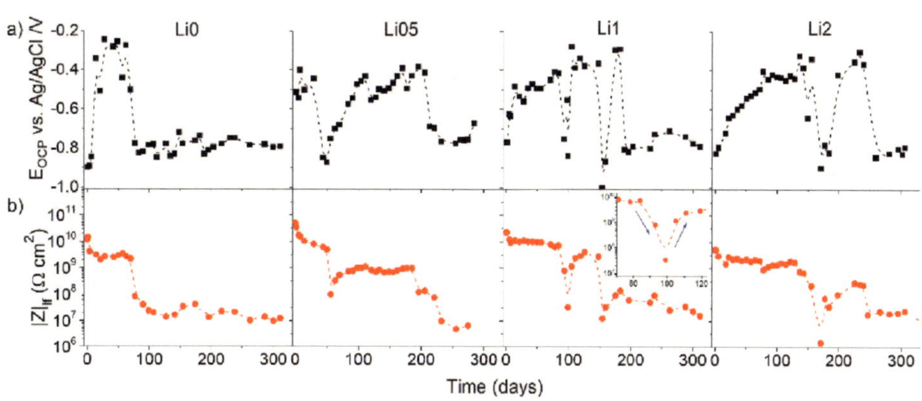

Figure 6.
(a) Time evolution of E_{OCP} and (b) $|Z_{lf}|$ for coatings modified with different amounts of lithium during 310 days of immersion in 3.5% NaCl solution. The inset shows in more detail the first impedance modulus recovery (99 days) for the Li1 sample (reprinted (adapted) with permission from [24]. Copyright (2020) American Chemical Society).

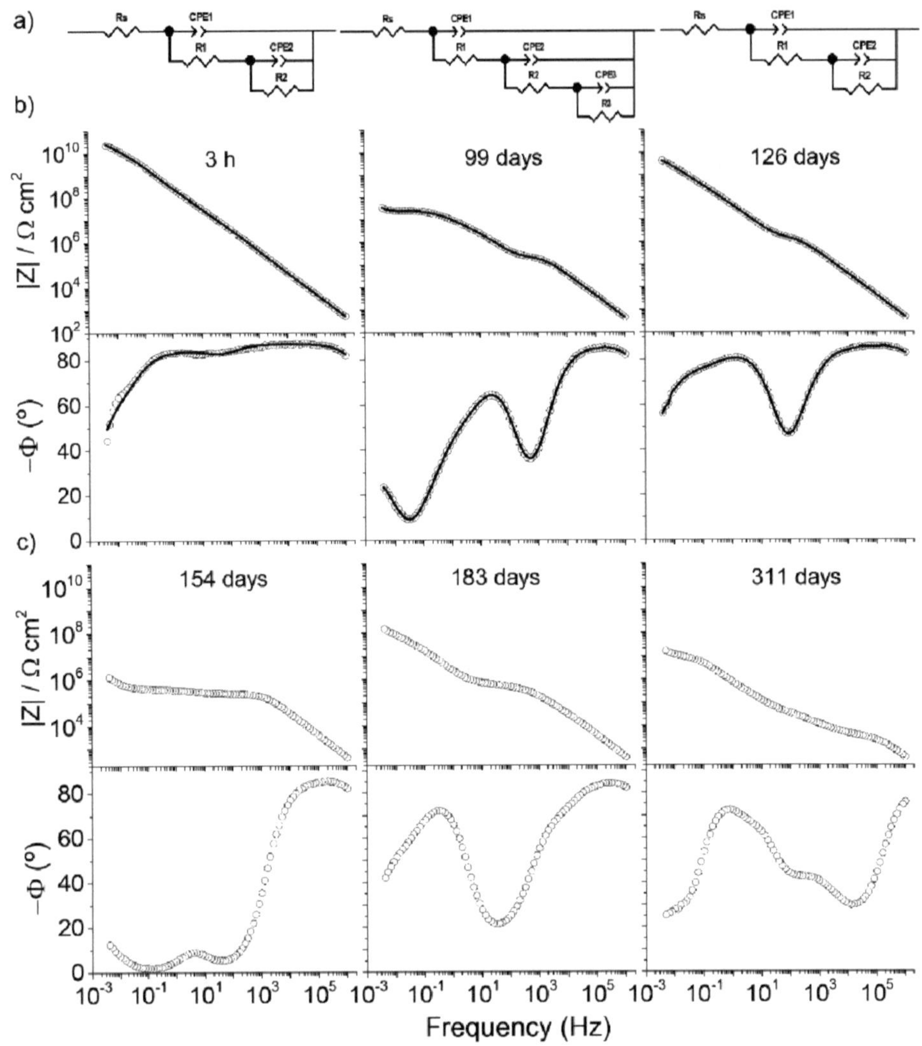

Figure 7.
(a) Electrical equivalent circuits used to fit (black lines) the EIS data of Li1 coating after (b) 3 h, 99 days, and 126 days of immersion in 3.5% NaCl solution and (c) time evolution of the bode plots after 154, 183, and 311 days of immersion (reprinted (adapted) with permission from [24]. Copyright (2020) American Chemical Society).

the first representing the upper water uptake layer and the second the inner intact/restored layer, and by adding a third time constant to simulate the corrosion process at the coating/metal interface (**Figure 7a**). Accordingly, the high-frequency data are generally attributed to phenomena occurring at the coating/electrolyte interface (R_1/CPE_1), the medium frequencies refer to the inner layer of the coating (R_2/CPE_2), and the low frequency is related to charge transfer resistance (R_{ct}) and double-layer capacitance (C_{dl}) at the coating/substrate interface (R_3/CPE_3). The electrochemical parameters, extracted by the fitting procedure, are shown in **Table 2**. After 99 days, a strong decay of R_1 and R_2 can be observed followed by a sharp increase after 126 days, from 0.20 MΩ cm^2 to 1.68 MΩ cm^2 and 0.02 MΩ cm^2 to 9.03 GΩ cm^2, respectively. As expected, an inverse behavior can be observed for the Q values of the CPE parameter, related to the coating capacitance.

To obtain more information on the Li-induced self-healing mechanism, surface analysis by time-of-flight secondary ion mass spectrometry (ToF-SIMS),

Li1	3 h		99 days		126 days	
χ^2	6.4×10^{-3}	—	4.5×10^{-3}	—	5.2×10^{-3}	—
R_s (Ω cm^2)	52.8	—	42.4	—	23.0	—
R_1 (MΩ cm^2)	17.1	(12.8)*	0.20	(2.0)	1.68	(2.0)
Q_1 (nΩ^{-1} cm^{-2} s^n)	0.46	(2.6)	0.55	(4.1)	0.66	(2.4)
n_1	0.97	(0.2)	0.97	(0.3)	0.95	(0.2)
R_2 (GΩ cm^2)	52.2	(3.8)	0.02	(1.1)	9.03	(3.6)
Q_2 (nΩ^{-1} cm^{-2} s^n)	0.21	(5.6)	17.0	(1.8)	3.85	(0.7)
n_2	0.70	(1.6)	0.77	(0.5)	0.88	(0.2)
R_3 (GΩ cm^2)			0.03	(14.6)		
Q_3 (nΩ^{-1} cm^{-2} s^n)			2723	(5.1)		
n_3			0.75	(4.9)		

**The values in brackets correspond to the error (%) of each parameter.*

Table 2.
Electrochemical parameters derived by fitting of the EIS data using electrical equivalent circuits of ***Figure 7a*** *for Li1 sample after 3 h, 99 days, and 126 days of immersion in NaCl 3.5%.*

X-ray photoelectron spectroscopy (XPS), and SEM were performed. Surface maps recorded by SIMS showed that the pit in the center of the immersion area (**Figure 8b**) contains only two small spots of lithium (red), surrounded by aluminum corrosion products (blue), and PMMA (green) containing a very low Li concentration (**Figure 8a**). In comparison, a higher Li surface concentration was detected outside the immersed area (**Figure 8c**), which indicates a Li^+ leaching process from the coating surface, evidencing its high mobility. Furthermore, the

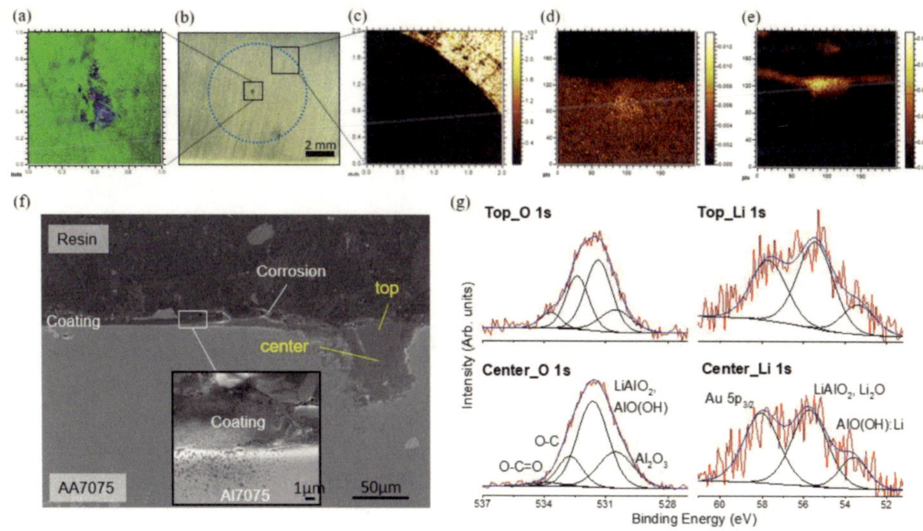

Figure 8.
(a) ToF-SIMS map of the Li1 coating overlay of Li^+ (red), $C_2H_3O_2^+$ (green), and Al^+ (blue); (b) optical micrograph displaying the immersed area delimited by a blue dashed circle; (c) normalized ToF-SIMS map of Li^+ (yellow) at the edge of the immersed zone; normalized ToF-SIMS maps of (d) $AlOH^+$ (yellow) and (e) Li^+ (yellow), recorded in top and center zones of ***Figure 8f****; (f) SEM cross-sectional view of the pit of the Li1 sample obtained after 310 days of immersion in 3.5% NaCl solution; and (g) fitted XPS O 1 s and Li 1 s spectra taken at the top and center of the pit. For SEM analysis, sputtered Au was used to improve surface conductivity (reprinted (adapted) with permission from [24]. Copyright (2020) American Chemical Society).*

cross-sectional image obtained by SEM (**Figure 8f**) shows a conical shape of the 70-μm-deep pit, surrounded by cathodic debris as a consequence of redox reactions. The cross-sectional SIMS maps of the pit revealed that its center is mainly filled with $AlOH_x$ corrosion products (**Figure 8d**) and that the top contains a lithium layer (**Figure 8c**), a clear evidence of the high mobility of lithium ions toward corrosive sites. XPS analysis performed in the top and center zones of the pit (**Figure 8f**) confirmed the SIMS findings, revealing a high Li concentration of 31 at.% and 22 at.%, respectively, and the presence of lithium oxide and lithium-containing aluminum oxide interphases (**Figure 8g**).

The reversible process observed for lithium-containing coatings is related to redox reactions taking place at the coating/substrate interface, as illustrated in **Figure 9**. As soon as water, oxygen, and chloride ions reach the substrate, Cl^- ions form a complex with aluminum producing soluble Al compounds, which cause the pit acidification/propagation. Simultaneously, at the top of the defect, $Al(OH)_x$ begins to precipitate at higher pH (reaction driven by hydroxyl ions, a product of carbonate action), leading to the formation of aluminum oxides. Finally, the curing activity of lithium within the pit is triggered at higher pH by the formation of Li^+ intercalated aluminum oxide phase with a highly passive character [16, 49, 50].

More rigorous tests of lithium self-healing activity were conducted by salt spray tests [24]. SEM images in **Figure 10a** and **b** show that after 7 days of testing, the Li2 sample was almost completely covered by a film, whereas the Li0 sample presents abundant corrosion products in the scratch track. EIS measurements performed after testing revealed for the Li2 coating an increase of the impedance modulus of one decade, while for the Li-free coating, a drop of impedance was observed. A confirmation of Li^+ activity in the corrosion zones was obtained by the overlay of Li^+ (red), $AlOH^+$ (green) and $C_2H_3O_2^+$ (yellow) SIMS map (**Figure 10c**) showing that lithium ions were preferentially leached from the walls of the scratch to active corrosion zones. The leached Li ions initiated a precipitation reaction of aluminum oxides, resulting in the formation of a protective layer within the scratch track that resulted in a significant reduction of the corrosion rate [46, 51].

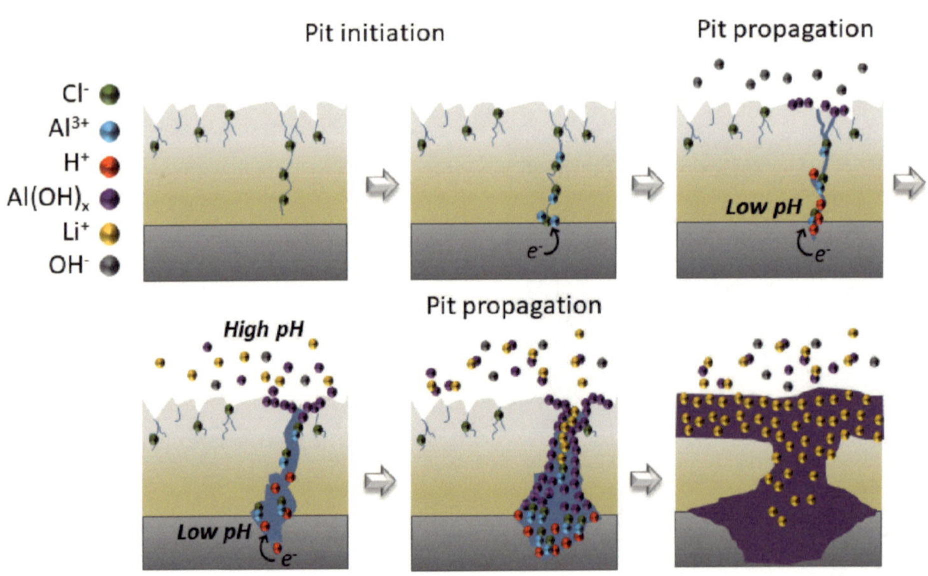

Figure 9.
Proposed mechanism for the formation of a lithium-induced protective barrier layer (reprinted (adapted) with permission from [24]. Copyright (2020) American Chemical Society).

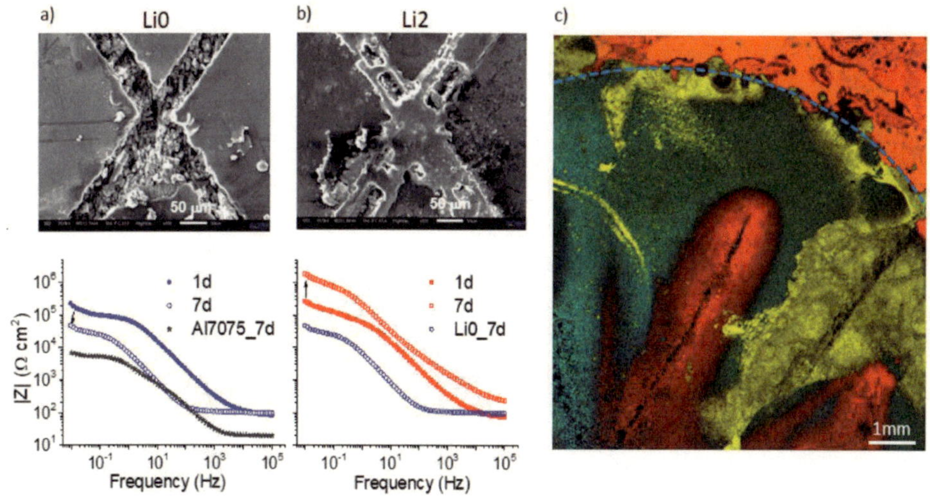

Figure 10.
SEM images of the scratched film recorded after 7 days of salt spray test and the corresponding EIS impedance modulus profiles after 1 day and 7 days for (a) Li0 and (b) Li2 coating on Al7075 substrate; (c) ToF-SIMS map of the Li2 sample showing the overlay map of Li^+ (red), $AlOH^+$ (green), and $C_2H_3O_2^+$ (yellow), after 7 days salt spray test (reprinted (adapted) with permission from [24]. Copyright (2020) American Chemical Society).

Structural, surface and electrochemical characterization showed that lithium ions were successfully incorporated into PMMA-silica coatings promoting a beneficial effect on the hybrid structure as well as smart corrosion inhibition. Based on the data of different surface analysis techniques, a self-healing mechanism was proposed that describes a lithium ion-induced formation of a protective layer of redox reaction products that block the corrosion process not only in localized defects (pits) but also in artificially damaged zones, thus extending considerably the service time of the PMMA-silica coating.

2.2.2 Corrosion inhibition by cerium ions

Recently, it was shown that organic-inorganic hybrids based on PMMA-cerium oxide are very promising materials for protective coating due to their excellent corrosion resistance, active self-healing ability, and environmental compatibility [29]. This nanocomposite material was prepared by the radical polymerization of methyl methacrylate and 2-hydroxyethyl methacrylate (HEMA), using benzoyl peroxide as thermal initiator, combined with the sol-gel synthesis of cerium oxide nanoparticles using $Ce(NO_3)_3{\cdot}6H_2O$ salt and LiOH. The hybrid solutions were used to deposit films on A1020 carbon steel by dip coating, yielding coatings with a thickness of 10 μm. Three coating formulations were evaluated with the following molar proportions, 1Ce:2HEMA:20MMA, 1Ce:2HEMA:25MMA, and 1Ce:2HEMA:30MMA, keeping the BPO/MMA molar ratio equal to 0.05. The HEMA molecule is formed by a methacrylate group that reacts with the organic phase (PMMA) and a hydroxyl terminal group that can be linked to the hydroxyl surface groups of the cerium oxide particles, thus acting as coupling agent and imparting excellent homogeneity and compatibility between the organic and inorganic phases (**Figure 11a**).

Structural characterization performed by XPS, SAXS, and high-resolution transmission electron microscopy (HRTEM) revealed the formation of CeO_2 and Ce_2O_3 nanoparticles with an average size of 2.5 nm and the homogenous distribution of these particles in the PMMA matrix through covalent bonds with the HEMA molecule (**Figure 11b**) [29]. The coatings deposited on carbon steel were transparent

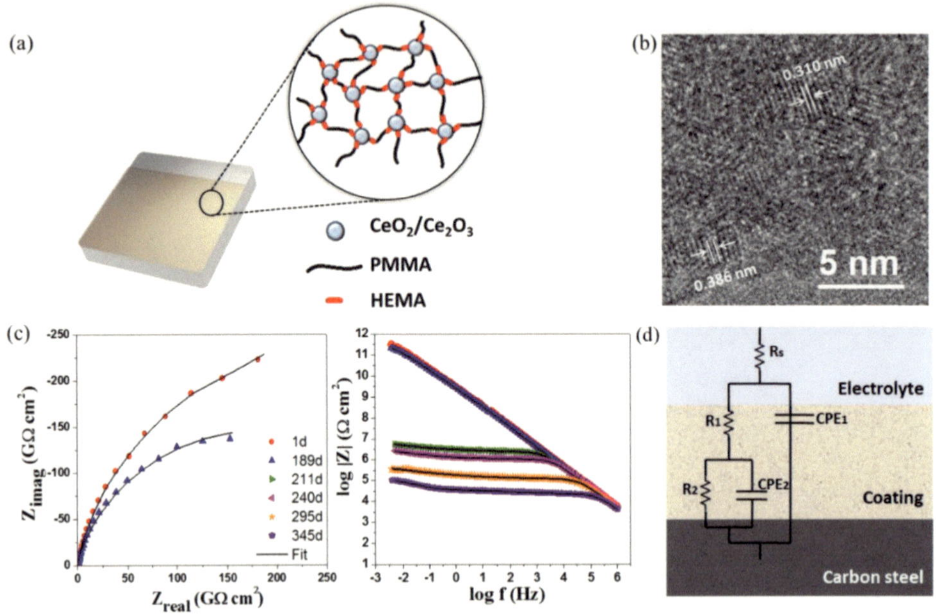

Figure 11.
(a) Schematic representation of the PMMA-cerium oxide coating deposited on carbon steel; (b) HRTEM image of the hybrid; (c) electrochemical impedance spectroscopy plots of the 1Ce:2HEMA:25MMA coating immersed in 3.5% NaCl solution; and (d) electrical equivalent circuit used to fit the EIS data (reproduced with permission from Elsevier [29]).

and free of pores and presented low surface roughness (<1.6 nm), extracted from AFM topography images. In addition, thermogravimetric analysis showed that the PMMA-cerium oxide hybrids have thermal stability up to 220°C and confirmed the nominal inorganic mass fraction for the samples, varying between 22 wt% (1Ce:2HEMA:20MMA) and 10 wt% (1Ce:2HEMA:30MMA).

Careful tuning of the molar ratio between the reagents yielded coatings with notable anticorrosive performance, assessed by electrochemical impedance spectroscopy in aggressive saline environment [29]. The 1Ce:2HEMA:20MMA sample exhibited low corrosion resistance, with impedance modulus at low frequency $|Z_{lf}|$ of 10 kΩ cm^2 after 1 day of immersion in 3.5% NaCl solution, only one order of magnitude higher than the uncoated carbon steel. A small increase of the polymeric content to 1Ce:2HEMA:25MMA, led to a sharp rise of the impedance modulus to 290 GΩ cm^2, which remained almost unchanged after more than 6 months in saline solution (**Figure 11c**). A further increase of the MMA amount to 1Ce:2HEMA:30MMA resulted in an intermediate performance of the coating, with $|Z_{lf}|$ of 95 GΩ cm^2 and durability of 5 months. The corrosion resistance achieved for the 1Ce:2HEMA:25MMA sample is comparable to that of the best anticorrosive coatings reported so far [22, 26, 27, 36, 43, 52–54], however, with the advantage of using nontoxic solvent or precursor. Several interesting results reported for high-performance organic-inorganic coatings with active corrosion protection are summarized in **Table 3**.

The electrical equivalent circuit shown in **Figure 11d** was used to fit the EIS data of the 1Ce:2HEMA:25MMA sample, allowing a deeper insight into the behavior of the electrochemical system. The circuit is composed of a solution resistance (R_s) in series with two time constants (R_1/CPE_1 and R_2/CPE_2) related to coating resistance and capacitance of the near-surface region and that of the inner layer close to the coating/steel interface, respectively. During the immersion period of 189 days,

Coating	Substrate	Thickness (μm)	$\lvert Z_{lf}\rvert$ (GΩ cm^2) 1st failure (days) solution	Ref.
PMMA-MPTS-TEOS-Ce(IV)	A1010 carbon steel	~2	~1, 304, 3.5% NaCl	[26]
PMMA-MPTS-TEOS-Li	AA7075	4–6	~1, 142, 3.5% NaCl	[24]
PMMA-HEMA-CeO_2	A1010 carbon steel	10	~290, 189, 3.5% NaCl	[29]
PMMA-MPTS-TEOS-Ce	Mild steel	26	~10, 362, 3.5% NaCl	[55]
Epoxy-HT-BZ	Carbon steel	35–45	~10, >60, 0.5 M NaCl	[56]
PVB-p-GAMo	Mild steel	47–53	~1, >21, 3.5% NaCl	[57]
Epoxy-MMT-Ce(III)	Carbon steel	50–70	~0.1, >100, 3.5% NaCl	[58]
Epoxy-PBH-GO	Carbon steel	—	~0.1, 40, 3.5% NaCl	[59]
Epoxy-APS-BS	AA2024	300	~1, >350, 0.5 M NaCl	[60]

PMMA, poly(methyl methacrylate); MPTS, 3-(trimethoxysilyl)propyl methacrylate; TEOS, tetraethoxysilane; Li, lithium; GO, graphene oxide; HT, hydrotalcites; BZ, benzoate; PVB, polyvinyl butyral; p-GAMo, porous organosilica with ion molybdate; MMT, montmorillonite;
PBH, polydopamine benzotriazole loaded halloysite nanotubes (HNTs); APS, (3-aminopropyl)trimethoxysilane; BS, bis[3-(triethoxysilyl)propyl]tetrasulfide.

Table 3.
Reported active hybrid coatings loaded with organic and/or inorganic corrosion inhibitors: composition, substrate, thickness, impedance modulus at low frequency $\lvert Z_{lf}\rvert$, time interval until the first failure event occurs, and solution.

the coating resistance of the bulk presented values up to 1 TΩ cm^2 and coating capacitance of less than 0.1 nF cm^{-2}, characteristics of an extremely efficient anticorrosion barrier arising from the dense and highly insulating cross-linked PMMA-cerium oxide structure [29].

Moreover, a detailed analysis of scratched and immersed coatings performed by EIS, XPS, and SEM (**Figure 12**) evidenced that Ce ions act as self-healing agents, by formation of insoluble cerium oxide and hydroxide species by reactions between leached cerium ions and hydroxyl groups in the scratch track, inhibiting the progression of the corrosion process and consequently enhancing the coating lifetime [29].

In summary, PMMA-cerium oxide coatings deposited on carbon steel combine high corrosion resistance, durability, self-healing property, low-cost, small-thickness, and eco-friendliness, thus representing a very promising alternative to conventional anticorrosive coatings for the protection of steel components.

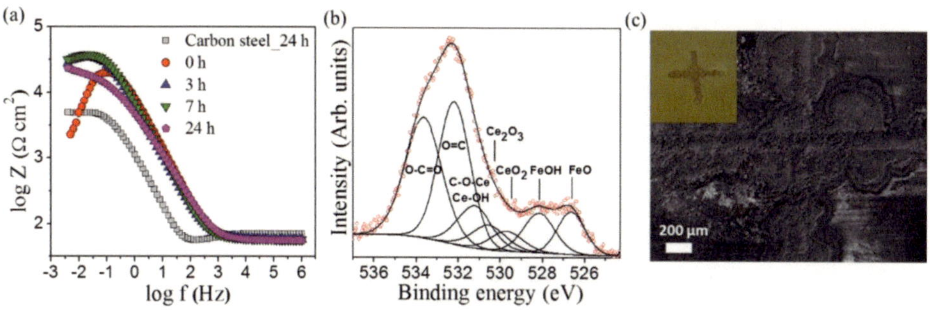

Figure 12.
(a) EIS plot of the scratched PMMA-cerium oxide coating immersed in 3.5% NaCl solution; (b) XPS O 1s spectra and (c) SEM image after immersion in 3.5% NaCl solution for 24 h (reproduced with permission from Elsevier [29]).

3. Conclusions

Advances in the search for effective substituents of chromates claim long-term protection through a dense barrier coating associated with an active response in case of damage. In this chapter, we have shown that acrylic nanocomposites efficiently prevent corrosion for 583 days by applying a thin (10 μm), adherent (up to 26 MPa), and transparent (80–90% transmission) layer of PMMA-silica on carbon steel. Furthermore, it was demonstrated that by incorporation of corrosion inhibitors, such as lithium ions and cerium oxide nanoparticles, a self-healing ability of hybrid coatings can be achieved, based on different mechanisms: lithium ions induce the formation of a passive layer in the corroded zone composed of Li^+ intercalated aluminum oxide phases, whereas cerium nanoparticles liberate cerium ions that react with water to form insoluble oxides and hydroxides in the affected zone, thus blocking the progress of the corrosion process of the AA7075 alloy and carbon steel, respectively. The efficient passive and active protection of metallic surfaces makes acrylic hybrid coatings potential candidates for a chromate-free future.

Acknowledgements

The authors would like to acknowledge the financial support of funding agencies, namely, Conselho Nacional de Desenvolvimento Científico e Tecnológico (CNPq) [grants 424133/2016-4, 307905/2018-7, 421081/2016-3], Coordenação de Aperfeiçoamento de Pessoal de Nível Superior (CAPES) [grants 465593/2014-3 and 88887.136401/2017-2100, Finance Code 001], and Fundação de Amparo à Pesquisa do Estado de São Paulo (FAPESP) [grants 2015/11907-2, 2015/09342-7, 2014/12182-9]. We also thank the National Laboratory of Synchrotron Light Source (LNLS) for SAXS beamline access and LNNano for the HRTEM images, the Hercules program under grant agreement ZW13_07 for ToF-SIMS measurements, and the Electrochemical and the Surface Engineering Group (SURF, Brussels/Belgium) for the SEM and EDX analysis.

Conflict of interest

The authors declare no conflict of interest.

Author details

Andressa Trentin, Samarah V. Harb, Thiago A.C. de Souza, Mayara C. Uvida and Peter Hammer*
Institute of Chemistry, São Paulo State University (UNESP), Araraquara, SP, Brazil

*Address all correspondence to: peter.hammer@unesp.br

IntechOpen

© 2020 The Author(s). Licensee IntechOpen. This chapter is distributed under the terms of the Creative Commons Attribution License (http://creativecommons.org/licenses/by/3.0), which permits unrestricted use, distribution, and reproduction in any medium, provided the original work is properly cited.

References

[1] Callister WD Jr, Rethwisch DG. Materials Science and Engineering: And Introduction. 9th ed. Hoboken: Wiley; 2014

[2] Lumley RN. Introduction: Aluminium, the strategic material. In: Lumley RN, editor. Fundamentals of Aluminium Metallurgy. Recent Advances. Duxford: Woodhead Publishing; 2018. pp. 17-30

[3] Harb SV, Trentin A, Torrico RFO, et al. Organic-inorganic hybrid coatings for corrosion protection of metallic surfaces. In: Giudice C, Canosa G, editors. New Technologies in Protective Coatings. Rijeka: IntechOpen; 2017. pp. 19-51

[4] Fischer HR, García SJ. Active protective coatings: Sense and heal concepts for organic coatings. In: Zheludkevich ML, Buchheit RG, Hughes AE, et al., editors. Active Protective Coatings: New-Generation Coatings for Metals. Dordrecht: Springer Science+Business Media B.V; 2016. pp. 139-156

[5] Hughes AE, Mol JMC, Zheludkevich ML, et al. Introduction. In: Hughes AE, Mol JMC, Zheludkevich ML, et al., editors. Active Protective Coatings: New-Generation Coatings for Metals. Dordrecht: Springer Science+Business Media B.V; 2016. pp. 1-13

[6] Fedrizzi L, Andreatta F. Corrosion inhibitors. In: Hughes RAE, Mol JMC, Zheludkevich ML, editors. Active Protective Coatings: New-Generation Coatings for Metals. Dordrecht: Springer Science+Business Media B.V; 2016. pp. 59-84

[7] IARC Working Group. Chromium (VI) compounds. In: Galichet L, editor. Arsenic, Metals, Fibres and Dusts. A Review of Human Carcinogens. International Agency for Research on Cancer; 2012. pp. 147-164

[8] Echeverría M, Abreu CM, Lau K, et al. Viability of epoxy-siloxane hybrid coatings for preventing steel corrosion. Progress in Organic Coatings. 2016;**92**:29-43

[9] Del Angel-López D, Domínguez-Crespo MA, Torres-Huerta AM, et al. Analysis of degradation process during the incorporation of ZrO_2:SiO_2 ceramic nanostructures into polyurethane coatings for the corrosion protection of carbon steel. Journal of Materials Science. 2013;**48**:1067-1084

[10] Harb SV, Trentin A, Uvida MC, et al. A comparative study on PMMA-TiO_2 and PMMA-ZrO_2 protective coatings. Progress in Organic Coatings. 2019;**140**:105477

[11] Trentin A, Gasparini AL, Faria FA, et al. Barrier properties of high performance PMMA-silica anticorrosion coatings. Progress in Organic Coatings. 2020;**138**:105398

[12] Dos Santos FC, Harb SV, Menu MJ, et al. On the structure of high performance anticorrosive PMMA-siloxane-silica hybrid coatings. RSC Advances. 2015;**5**:106754-106763

[13] Gandhi JS, Singh S, Van Ooij WJ, et al. Evidence for formation of metallo-siloxane bonds by comparison of dip-coated and electrodeposited silane films. Journal of Adhesion Science and Technology. 2006;**20**:1741-1768

[14] Torrico RFAO, Harb SV, Trentin A, et al. Structure and properties of epoxy-siloxane-silica nanocomposite coatings for corrosion protection. Journal of Colloid and Interface Science. 2018;**513**:617-628

[15] Sorensen PA, Kiil S, Dam-Johansen K, et al. Anticorrosive

coatings: A review. Journal of Coatings Technology and Research. 2009;**6**:135-176

[16] Grundmeier G, Schmidt W, Stratmann M. Corrosion protection by organic coatings: Electrochemical mechanism and novel methods of investigation. Electrochimica Acta. 2000;**45**:2512-2533

[17] Sanchez C, Belleville P, Popall M, et al. Applications of advanced hybrid organic-inorganic nanomaterials: From laboratory to market. Chemical Society Reviews. 2011;**40**:696-753

[18] Alessi S, Toscano A, Pitarresi G, et al. Water diffusion and swelling stresses in ionizing radiation cured epoxy matrices. Polymer Degradation and Stability. 2017;**144**:137-145

[19] Blaiszik BJ, Kramer SLB, Olugebefola SC, et al. Self-healing polymers and composites second Sandia fracture challenge view project shock wave energy dissipation by mechanochemically-active materials view project self-healing polymers and composites. Annual Review of Materials Research. 2010;**40**:179-211

[20] Qiu S, Chen C, Zheng W, et al. Long-term corrosion protection of mild steel by epoxy coating containing self-doped polyaniline nanofiber. Synthetic Metals. 2017;**229**:39-46

[21] Maya-Visuet E, Gao T, Soucek M, et al. The effect of TiO_2 as a pigment in a polyurethane/polysiloxane hybrid coating/aluminum interface based on damage evolution. Progress in Organic Coatings. 2015;**83**:36-46

[22] Ammar S, Ramesh K, Vengadaesvaran B, et al. Amelioration of anticorrosion and hydrophobic properties of epoxy/PDMS composite coatings containing nano ZnO particles. Progress in Organic Coatings. 2016;**92**:54-65

[23] Visser P, Marcoen K, Trindade GF, et al. The chemical throwing power of lithium-based inhibitors from organic coatings on AA2024-T3. Corrosion Science. 2019;**150**:194-206

[24] Trentin A, Harb SV, Uvida MC, et al. Dual role of lithium on the structure and self-healing ability of PMMA-silica coatings on AA7075 alloy. ACS Applied Materials & Interfaces. 2019;**11**:40629-40641

[25] Lv X, Li X, Li N, et al. ZrO_2 nanoparticle encapsulation of graphene microsheets for enhancing anticorrosion performance of epoxy coatings. Surf and Coatings Technology. 2019;**358**:443-451

[26] Harb SV, Dos Santos FC, Caetano BL, et al. Structural properties of cerium doped siloxane-PMMA hybrid coatings with high anticorrosive performance. RSC Advances. 2015;**5**:15414-15424

[27] Calado LM, Taryba MG, Carmezim MJ, et al. Self-healing ceria-modified coating for corrosion protection of AZ31 magnesium alloy. Corrosion Science. 2018;**142**:12-21

[28] Schem M, Schmidt T, Gerwann J, et al. CeO_2-filled sol-gel coatings for corrosion protection of AA2024-T3 aluminium alloy. Corrosion Science. 2009;**51**:2304-2315

[29] Harb SV, Trentin A, de Souza TAC, et al. Effective corrosion protection by eco-friendly self-healing PMMA-cerium oxide coatings. Chemical Engineering Journal. 2019;**383**:123219

[30] Chen C, Qiu S, Cui M, et al. Achieving high performance corrosion and wear resistant epoxy coatings via incorporation of noncovalent functionalized graphene. Carbon. 2017;**114**:356-366

[31] Xia Z, Liu G, Dong Y, et al. Anticorrosive epoxy coatings based on

polydopamine modified molybdenum disulfide. Progress in Organic Coatings. 2019;**133**:154-160

[32] Khodabakhshi J, Mahdavi H, Najafi F. Investigation of viscoelastic and active corrosion protection properties of inhibitor modified silica nanoparticles/epoxy nanocomposite coatings on carbon steel. Corrosion Science. 2019;**147**:128-140

[33] Nardeli JV, Fugivara CS, Taryba M, et al. Tannin: A natural corrosion inhibitor for aluminum alloys. Progress in Organic Coatings. 2019;**135**:368-381

[34] Brinker CJ, Scherer GW. Sol → gel → glass: I. Gelation and gel structure. Journal of Non-Crystalline Solids. 1985;**70**:301-322

[35] Lou C, Zhang R, Lu X, et al. Facile fabrication of epoxy/polybenzoxazine based superhydrophobic coating with enhanced corrosion resistance and high thermal stability. Colloids Surfaces A. 2019;**562**:8-15

[36] Lamaka SV, Xue HB, Meis NNAH, et al. Fault-tolerant hybrid epoxy-silane coating for corrosion protection of magnesium alloy AZ31. Progress in Organic Coatings. 2015;**80**:98-105

[37] He P, Wang J, Lu F, et al. Synergistic effect of polyaniline grafted basalt plates for enhanced corrosion protective performance of epoxy coatings. Progress in Organic Coatings. 2017;**110**:1-9

[38] Díaz I, Chico B, De La Fuente D, et al. Corrosion resistance of new epoxy-siloxane hybrid coatings. A laboratory study. Progress in Organic Coatings. 2010;**69**:278-286

[39] Ashhari S, Sarabi AA, Kasiriha SM, et al. Aliphatic polyurethane-montmorillonite nanocomposite coatings: Preparation, characterization, and anticorrosive properties. Journal of Applied Polymer Science. 2011;**119**:523-529

[40] Lu F, Song B, Wang J. Preparation and characterization of acrylic polyurethane/polyaniline nanocomposite coatings. International Journal of Electrochemical Science. 2017;**12**:2679-2691

[41] Hammer P, Dos Santos FC, Cerrutti BM, et al. Highly corrosion resistant siloxane-polymethyl methacrylate hybrid coatings. Journal of Sol-Gel Science and Technology. 2012;**63**:266-274

[42] Harb SV, Pulcinelli SH, Santilli CV, et al. A comparative study on graphene oxide and carbon nanotube reinforcement of PMMA-siloxane-silica anticorrosive coatings. ACS Applied Materials & Interfaces. 2016;**8**:16339-16350

[43] Ammar S, Ramesh K, Vengadaesvaran B, et al. A novel coating material that uses nano-sized SiO_2 particles to intensify hydrophobicity and corrosion protection properties. Electrochimica Acta. 2016;**220**:417-426

[44] Alrashed MM, Jana S, Soucek MD. Corrosion performance of polyurethane hybrid coatings with encapsulated inhibitor. Progress in Organic Coatings. 2019;**130**:235-243

[45] He Z, Li X, Soucek MD, et al. Inhibition of acid undercutting of inorganic/organic hybrid polyurethane coatings. Progress in Organic Coatings. 2019;**134**:169-176

[46] Visser P, Lutz A, Mol JMC, et al. Study of the formation of a protective layer in a defect from lithium-leaching organic coatings. Progress in Organic Coatings. 2016;**99**:80-90

[47] Guinier G, Fournet C, Walker CB, et al. Small Angle Scattering of Xrays. New York: Freeman; 1955

[48] Hammouda B. A new Guinier-Porod model. Journal of Applied Crystallography. 2010;**43**:716-719

[49] Foley RT. Localized corrosion of aluminum alloys—A review. Corrosion. 1986;**42**:277-288

[50] Reboul MC, Warner T, Mayet H, et al. A ten-step mechanism for the pitting corrosion of aluminium. Materials Science Forum. 1996;**217-222**:1553-1558

[51] Isupov VP. Intercalation compounds of aluminum hydroxide. Journal of Structural Chemistry. 1999;**40**:672-682

[52] Thai TT, Druart M-E, Paint Y, et al. Influence of the sol-gel mesoporosity on the corrosion protection given by an epoxy primer applied on aluminum alloy 2024-T3. Progress in Organic Coatings. 2018;**121**:53-63

[53] Wang N, Gao H, Zhang J, et al. Effect of graphene oxide/ZSM-5 hybrid on corrosion resistance of waterborne epoxy coating. Coatings. 2018;**8**:1-14

[54] Ramezanzadeh B, Bahlakeh G, Ramezanzadeh M. Polyaniline-cerium oxide (PAni-CeO_2) coated graphene oxide for enhancement of epoxy coating corrosion protection performance on mild steel. Corrosion Science. 2018;**137**:111-126

[55] Mosa J, Rosero-Navarro NC, Aparicio M. Active corrosion inhibition of mild steel by environmentally-friendly Ce-doped organic-inorganic sol-gel coatings. RSC Advances. 2016;**6**:39577-39586

[56] Nguyen Thuy D, To Thi Xuan H, Nicolay A, et al. Corrosion protection of carbon steel by solvent free epoxy coating containing hydrotalcites intercalated with different organic corrosion inhibitors. Progress in Organic Coatings. 2016;**101**:331-341

[57] Lin Y, Chen R, Zhang Y, et al. Application of porous organosilica intercalated graphene oxide as the container of molybdate inhibitors for anticorrosive coatings. Materials and Design. 2020;**186**:108304

[58] Danaee I, Darmiani E, Rashed GR, et al. Self-healing and anticorrosive properties of Ce(III)/Ce(IV) in nanoclay-epoxy coatings. Iranian Polymer Journal. 2014;**23**:891-898

[59] Chen C, Xiao G, He Y, et al. Bio-inspired superior barrier self-healing coating: Self-assembly of graphene oxide and polydopamine-coated halloysite nanotubes for enhancing corrosion resistance of waterborne epoxy coating. Progress in Organic Coatings. 2019;**139**:105402

[60] Abdolah Zadeh M, Van Der Zwaag S, Garcia SJ. Adhesion and long-term barrier restoration of intrinsic self-healing hybrid sol-gel coatings. ACS Applied Materials & Interfaces. 2016;**8**:4126-4136

Section 3

General Corrosion

Chapter 5

The Arrhenius Acid and Base Theory

Shikha Munjal and Aakash Singh

Abstract

Swedish Svante Arrhenius, in 1884 proposed the concept of acid and base based on the theory of ionization. According to Arrhenius, the acids are the hydrogen-containing compounds which give H^+ ions or protons on dissociation in water and bases are the hydroxide compounds which give OH^- ions on dissociation in water. This concept is only applicable to those compounds which dissolved in aqueous solution (or you can say where water is the solvent). It covers many common acids, bases and their chemical reactions, but there are also other compounds that have the characteristics of acids and bases but they do not fit into Arrhenius concept.

Keywords: acids and bases, Arrhenius acid-base theory, the water-ion system, amphoteric nature of water, hydrogen or hydronium ion

1. Introduction

The concept of acids and bases have been defined many times in different ways. Several scientists put various definitions to characterize the acids and bases in which some of the concepts are quite narrow and some are comprehensive. Acids and bases are existing everywhere in our daily life. Every liquid except water, that we used having acid and basic properties, for example, vinegar (contains acetic acid), soft drinks (contains carbonic acid), buttermilk (contains lactic acid), soap (contains base). The earliest definitions were made on the basis of their taste and their effect on other substances.

1.1 Acids

Acid are those substances which have sour in taste, sharp odor, corrosive, having pH < 7 and turn blue litmus red. The neutralization reaction occurs when acid reacts with alkali, forms salt and water. The products are less acidic or basic than reactants. It reacts with metals, produces H_2. For example, the reaction between sodium hydroxide (base) and hydrochloric acid, forms sodium chloride (salt) and water.

$$NaOH\ (aq) + HCl\ (aq) \xrightarrow{\text{Neutralization}} NaCl\ (aq) + H_2O\ (l)$$

1.2 Factors affecting acidic strength

The strength of acids and bases depends on following factors:

- Polarity of the molecule and strength of H–A bond
- Electro negativity
- Size

1.2.1 Polarity of the molecule and strength of H–A bond

As the polarity of the molecule increases, the electron density will get away from hydrogen atom and it becomes H^+ (proton). The greater is the positive charge on the hydrogen atom, H–A bond will become weaker, lesser is the energy required to break it. Then, the proton will easily dissociate in the solution. Hence, it will be the strong acid [1].

$$H - A \xrightarrow{\text{water}} H_3O^+ + A^- \text{ (A is more electronegative than H)}$$

1.2.1.1 Key points

The priority should be given to the polarity of H–A bond, when we compare the acidic strength of elements in the same row. But when we compare the acidic strength of elements of same group of periodic table, then priority is given to strength of H–A bond.

1.2.2 Electro negativity

The hydrogen is attach to the more electronegative atom is more acidic. For example - the hydrogen is attached to the oxygen (E.N = 3.5) is more acidic than the hydrogen attached to nitrogen (E.N = 3.0, which is less electronegative than oxygen).

1.2.3 Size

The size of "A" atom affects the acidity of acidic strength. As the size of the atom increases, the bond becomes weaker and acidic strength increases [2].

1.2.4 Hybridization

Hybridization plays an important role in determining the acidic strength. As the hybridization moves from sp^3 to sp, the "s" character increases, so acidic strength increases.

1.3 Bases

Bases are those substances which have bitter taste, odorless, turn red litmus blue, having pH more than 7 and becomes less alkaline when react with acid. These are violent and less reactive than acids. For example, NaOH (Sodium hydroxide), LiOH (Lithium hydroxide), KOH (potassium hydroxide), etc.

These are the general properties of acids or bases, but not true for every single acid or base. There are some important concepts:

- Arrhenius acid-base theory
- Lewis acid-base concept

- Bronsted-Lowry concept
- Lux-flood concept
- Solvent-system concept

2. The Arrhenius acid-base theory (the water-ion system)

The Arrhenius acid-base theory was proposed by Swedish Svante Arrhenius. It was the first modern approach to acid-base concept. This theory is quite simple and useful. According to Arrhenius theory, acids are the compound that increases the concentration of H^+ or proton in aqueous solution. The released H^+ ion or proton is not free-floating proton, it exists in combined state with the water molecule and forms hydronium ion (H_3O^+). The common examples of Arrhenius acid includes HCl (hydrochloric acid), H_2SO_4 (sulphuric acid), HNO_3 (nitric acid), etc. as shown in **Table 1**.

$$HCl\,(aq) \longrightarrow H^+\,(aq) + Cl^-\,(aq)$$

When it is dissolved in water, then:

$$HCl\,(aq) \longrightarrow H_3O^+\,(aq) + Cl^-\,(aq)$$

$$HNO_3\,(aq) \xrightarrow{\text{Water}} H_3O^+\,(aq) + NO_3^-\,(aq)$$

The acids like HNO_3, HCl, etc. gives one proton on dissociation, called monoprotic acids. The acids like H_2SO_4, H_3PO_4, etc. which having more than one hydrogen atoms and gives more than 1 H^+ ions on dissociation, called polyprotic acids. It is not necessary that polyprotic acids are stronger than monoprotic acids.

$$H_2SO_4(aq) \xrightarrow{\text{Water}} H^+\,(aq) + HSO_4^-\,(aq)$$

$$HSO_4^-(aq) \xrightarrow{\text{Water}} H^+\,(aq) + SO_4^{2-}\,(aq)$$

$$\text{or}\quad 2H_2SO_4\,(aq) \xrightarrow{\text{Water}} 2H^+\,(aq) + SO_4^{2-}\,(aq)$$

Similarly, Arrhenius bases are compounds that increase the concentration of OH^- or hydroxide ion in aqueous solution or having at least one OH^- ion in

Arrhenius acid formula	Name
$HClO_3$	Chloric acid
HNO_3	Nitric acid
$HClO_4$	Perchloric acid
H_3PO_4	Phosphoric acid
H_2SO_4	Sulphuric acid
H_2SO_3	Sulfurous acid
HCl	Hydrochloric acid
CH_3COOH	Acetic acid
HBr	Hydrobromic acid

Table 1.
Some Arrhenius acids.

formula. The common examples of Arrhenius base includes NaOH (sodium hydroxide), KOH (potassium hydroxide), $Ca(OH)_2$ (calcium hydroxide), $Mg(OH)_2$ (magnesium hydroxide), NH_4OH (ammonium hydroxide), etc. as shown in **Table 2**.

When sodium hydroxide dissolved in water, it fully dissociates into ions Na^+ and OH^-, this dissociation increases the concentration of hydroxide ions in the solution.

$$NaOH\ (aq) \xrightarrow{Water} Na^+\ (aq) + OH^-\ (aq)$$

$$KOH\ (aq) \xrightarrow{Water} K^+\ (aq) + OH^-\ (aq)$$

$$Ca(OH)_2\ (aq) \xrightarrow{Water} Ca^{2+}\ (aq) + 2OH^-\ (aq)$$

$$Mg(OH)_2\ (aq) \xrightarrow{Water} Mg^{2+}\ (aq) + 2OH^-(aq)$$

$$NH_4OH\ (aq) \xrightarrow{Water} NH_4\ (aq) + OH^-\ (aq)$$

2.1 Neutralization reaction

When Arrhenius acid and Arrhenius base reacts, salt and water is formed as product, the reaction is known as neutralization reaction. For example:

$$HCl + NaOH \xrightarrow{Neutralisation} NaCl + H_2O$$

The acids which are completely ionized in aqueous solution, is termed as strong acids such as HCl, HNO_3, H_2SO_4, etc.

$$HCl\ (aq) \xrightarrow{Water} H_3O^+ + Cl^-$$

Hydrochloric acid is a strong acid. When it dissociates into water, hydronium ion and chloride ions are formed as product. Chloride ions are weak base, but its basicity does not make the solution basic because acidity is overpowering the basicity of chloride ions. The H^+ ions combine with water molecule and form hydronium ion. In case of strong acid, the concentration of hydronium ion formed is equal to the concentration of the acid whereas in case of weak acids, the concentration of hydronium ions in solution is always less than the concentration of hydrogen ions.

Arrhenius base	**Name**
NaOH	Sodium hydroxide
NH_4OH	Ammonium hydroxide
KOH	Potassium hydroxide
$Mg(OH)_2$	Magnesium hydroxide
$Ca(OH)_2$	Calcium hydroxide
$Al(OH)_3$	Aluminum hydroxide

Table 2.
Some Arrhenius bases.

Whereas the acids which are weakly ionized in aqueous solution, is termed as weak acids such as acetic acid (CH_3COOH).

$$CH_3COOH\ (aq) \xrightarrow{\text{Water}} CH_3COO^- + H^+$$

In case of weak acids, the concentration of hydronium ion is always less than the concentration of acid.

Similarly, bases which are completely ionized in aqueous solution, are termed as strong bases such as NaOH, KOH, etc. whereas the bases which are weakly ionized in aqueous solution, is known as weak bases such as ammonium hydroxide (NH_4OH), calcium hydroxide ($Ca(OH)_2$), etc.

Note: It is not necessary that strong acids/bases are concentrated and weak acids/bases are dilute. Because, the dissociation of a substance does not depend on its concentration.

3. Utility of Arrhenius concept

This theory explains many phenomena like strength of acids and bases, salt hydrolysis and neutralization.

4. Hydrogen ion (H^+) or hydronium ion (H_3O^+)

When electron is removing from hydrogen atom, hydrogen ion H^+ is formed which is very reactive. But this H^+ ion does not exist in aqueous solution. Since in aqueous medium, it reacts with water molecule and forms hydronium ion (H_3O^+). Water is a polar molecule; it has the ability to attract the hydrogen ion (H^+). The water contains hydrogen and oxygen in which oxygen (EN = 3.5) is more electronegative that pulls the electron density towards it and causing the partial negative charge on the molecule. Due to partial negative charge, it has ability to attract the positively charged hydrogen ion (H^+) and form hydronium ion (H_3O^+). Hydronium ions are more stable than hydrogen ions.

$$H_2O\ (l) \rightleftharpoons H^+\ (aq) + OH^-\ (aq)$$

The hydronium ion is very important factor in chemical reaction that occurs in aqueous solutions [3]. It is formed by the protonation of water.

5. Concept of pH

The pH of solution can be determined by the concentration of hydronium ion.

$$pH = -\log\ (H_3O^+)$$

From this equation, we can find the pH of pure water. The pH of pure water comes to be 7 that is considered to be neutral. The solution is either acidic or basic depending on the change in the concentration of hydronium ion.

According to **Figure 1**:

- If the concentration of the hydronium ion in the solution increases means more than 10^{-7} mol/l, pH increases that makes the solution more acidic.

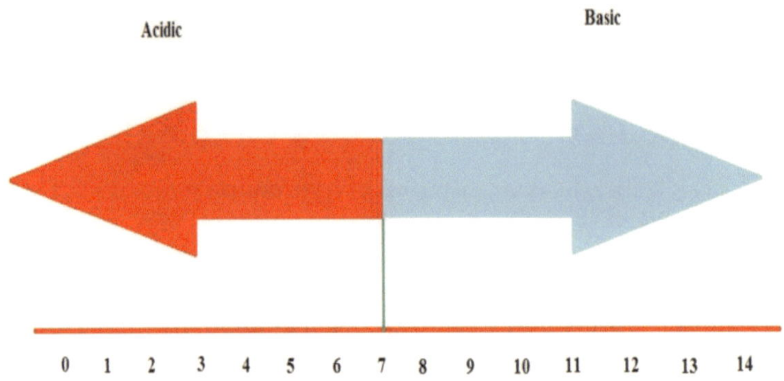

Figure 1.
pH scale.

- If the concentration of the hydronium ion in the solution decreases means $<10^{-7}$ mol/l, pH decreases that makes the solution more basic.

6. Amphoteric nature of water

The word amphoteric is derived from Greek word "amphi" that means both (acid and base). Amphoteric substances are those that has potential to act either as an acid or base. For example: H_2O (water) [3].

$$H_2O\,(l) \rightleftharpoons H^+\,(aq) + OH^-\,(aq)$$

On dissociation, it ionizes into H^+ and OH^- (hydroxide) ion. The presence of H^+ indicates an acid and the presence of OH^- ion indicates a base. Since, water is a neutral molecule. So, it dissociates equally into H^+ and OH^- ion.

According to Arrhenius acid-base theory:

The amphoteric nature of water is very important because most of the acid-base chemical reactions takes place in the presence of water. Water is important amphoteric compound that can act as both an Arrhenius acid or Arrhenius base.

Auto-ionization of water [4];

$$H_2O \overset{H_2O}{\rightleftharpoons} H_3O^+\,(aq) + OH^-\,(aq)$$

The H^+ ion (a bare proton) does not exist in the solution, it forms hydronium ions by hydrogen bonding with nearest water molecule. Many books refers the "concentration of hydrogen ions" which is not correct. Because there are no H^+ ions, only hydronium ions in the solution. Technically, the number of hydronium ions formed is equal to the number of hydrogen ion. So, both can be used.

7. Advantages of Arrhenius theory

This theory is used to explains:

- Strength of acid and bases

The strength of Arrhenius acid and Arrhenius base can be determined by the extent to which it dissociate to give H^+ ion or hydroxide ion [5].

- The properties of acids and bases in aqueous medium.
- Neutralization of acid by reaction with base

8. Limitation of Arrhenius acid-base theory

1. This theory is very limited, out of three theories. According to this theory, the solution medium should be aqueous and acid should produce hydrogen ion (H^+) or base should produce hydroxide ion (OH^-) on dissociation with water. Hence, the substance is regarded as Arrhenius acid or Arrhenius base when it is dissolved in water. For example, HNO_3 is regarded as Arrhenius acid when it is dissolved in aqueous solution. But when it is dissolved in any other solvent like benzene, no dissociation occurs. This is against the Arrhenius theory.

2. Arrhenius theory is not applicable on the non-aqueous or gaseous reactions because it explained the acid-base behavior in terms of aqueous solutions.

3. In Arrhenius theory, salts are produce in the product which are neither acidic nor basic. So, this theory cannot explain the neutralization reaction without the presence of ions. For example, when acetic acid (weak acid) and sodium hydroxide (strong base) reacts, then the resulting solution basic. But this concept is not explained by Arrhenius.

4. Arrhenius theory is only applicable to those compounds which having formula HA or BOH for acids and bases. There are some acids like $AlCl_3$, $CuSO_4$, CO_2, SO_2 which cannot be represented by HA formula, this theory is unable to explain their acidic behavior. Similarly, there are some bases like Na_2CO_3, NH_3, etc. which do not represented by BOH formula, this theory is unable to explain their basic behavior.

9. Bronsted-Lowry theory

We have been previously learned an Arrhenius acid-base theory which provided a good start towards the acid-base chemistry but it has certain limitations and problems. After this theory, a Danish chemist, named Johannes Nicolaus Bronsted and British scientist, Thomas Martin Lowry proposed a different definition of acid-base that based on the abilities of compound to either donate or accept the protons. This theory is known as Bronsted-Lowry theory, also called Proton theory of acid and base. This theory gives a more general and useful acid-base definition and applies to wide range of chemical reactions. In this theory, we usually consider a hydrogen atom as a proton that has lost its electrons and becomes a positively charged hydrogen ion (represented by symbol, H^+).

According to Bronsted-Lowry concept, an acid is considered to be Bronsted-Lowry acid which is capable to donate a proton to someone else. A base is considered to be Bronsted-Lowry base which is capable to accept a proton from someone else. From here, it can note that when an acid reacts with a base, the proton is transferred from one chemical species to another.

- Bronsted-Lowry acid-Hydrogen ion (Proton) donor.
- Bronsted-Lowry base-Hydrogen ion (Proton) acceptor.

9.1 Conjugate acid-base pairs

- Conjugate acid: It forms when a base accepts proton.
- Conjugate base: It forms when an acid donates proton.

Note: If an acid is strong, the conjugate base will be weaker and if the base is strong, the conjugate acid will be weak.

Consider the following chemical reaction:

$$HCl\ (aq) + NH_3\ (aq) \rightleftarrows NH_4^+\ (aq) + Cl^-\ (aq)$$

Hydrogen Chloride Ammonia Ammoniumion Chlorideion

In this reaction, HCl is an acid because it is donating proton to NH_3. Therefore, HCl is act as Bronsted-Lowry acid whereas NH_3 has a lone pair of electrons which is used to accept the protons. Therefore, NH_3 is act as Bronsted-Lowry base. This reaction is reversible also. In reversible case, the ammonium ion reacts with chloride ion and again converts into ammonia (NH_3) and hydrogen chloride (HCl). In this case, the ammonium ion is donating a proton, called conjugate acid. The chloride ion (Cl^-) ion is accepting a proton, called conjugate base.

There are two conjugate pairs—conjugate pair 1 and conjugate pair 2.

Conjugate pair 1: HCl and Cl^-

Conjugate pair 2: NH_3 and NH_4^+

From that equation, the ammonium ion (NH_4^+) is a conjugate acid of base ammonia and chloride ion (Cl^-) is a conjugate base of acid hydrogen chloride.

Note: According to the theory of Arrhenius, the reaction between HCl and NH_3 is not considered as acid-base reaction because none of these species gives H^+ and OH^- ions in water.

9.2 Examples of Bronsted-Lowry acids and bases

$$HNO_3\ (aq) + H_2O\ (l) \longrightarrow H_3O^+\ (aq) + NO_3^-\ (aq)$$

In this reaction, the nitric acid donates a proton to the water, therefore it act as a Bronsted-Lowry acid. Since, water accepts a proton from nitric acid, so it is act as Bronsted-Lowry base. In this reaction, the arrow is drawn only to the right side which means that reaction highly favours the formation of products.

$$NH_3\ (aq) + H_2O\ (l) \longrightarrow NH_4^+\ (aq) + OH^-\ (aq)$$

In this reaction, the water is losing its proton, becomes hydroxide (OH^-) and donates its proton to the ammonia. Therefore, water is act as Bronsted-Lowry acid. Ammonia is accepting a proton from the water, forms ammonium ion (NH_4^+). Therefore, ammonia is act as Bronsted-Lowry base.

From the above two reactions, we can conclude that the water is Amphoteric in nature which means that it can act as both: Bronsted-Lowry acid and Bronsted-Lowry base.

9.3 Advantages of Bronsted-Lowry theory

1. This theory is able to explain the acid-base behavior in aqueous and non-aqueous medium.

2. It explains the basic character of substances like NH_3, CaO, Na_2CO_3, that is, which do not contain –OH group but according to Arrhenius theory, they are not considered as bases.

3. It explains the acidic character of substances like CO_2, SO_2, etc. which do not contain hydrogen ion group but according to Arrhenius theory, they are not considered as acids.

4. This theory also explains the acid-base behavior of ionic species.

9.4 Disadvantages of Bronsted-Lowry theory

1. According to Bronsted-Lowry theory, same compound is act as acid in one reaction and act as base in other reaction. So, sometimes it is very difficult to predict the exact acid or base in a reaction.

2. This theory is not able to explain the acidic, basic as well as Amphoteric gaseous molecule.

3. This theory does not explain the behavior of acids like BF_3, $AlCl_3$, BCl_3, etc. which do not protons to loose or donate.

4. This theory does not explain reactions between acidic oxides (CO_2, SO_2, SO_3) and basic oxides (CaO, MgO, BrO) which takes place in the absence of solvent.

10. Relation between Arrhenius theory and Bronsted-Lowry theory

These two theories are not against to each other in any way, in fact Bronsted-Lowry theory is advance to the Arrhenius theory.

According to the Arrhenius theory, a substance which produces hydrogen ion in water, called acid. A substance which produces hydroxide ion in water, called base.

According to Bronsted-Lowry theory, an acid is proton donor and base is proton acceptor.

$$HCl\ (g) + H_2O\ (l) \longrightarrow H_3O+\ (aq) + Cl\text{-}\ (aq)$$

According to Arrhenius theory, hydrochloric acid is an acid which gives hydrogen ions in water but according to Bronsted-Lowry theory, hydrochloric acid is an acid because it donates a proton to the water molecule. By observing both concepts, water is acting as a base. So, we can see here that both theories are very similar to each other.

Actually, Arrhenius theory is limited only to the aqueous solution. It does not explain the acid behavior in gaseous form.

11. Lewis acid-base theory

The Bronsted-Lowry theory which we have been previously studied was a good startup for acid-base chemistry. The Bronsted-Lowry concept was based on the transfer of proton from one chemical species to another. But this theory has certain limitations. UC Berkeley scientist, G.N. Lewis, in 1923 proposed a new acid-base theory which is based on their transfer of electrons. This theory is more advanced and flexible than Bronsted-Lowry because it explains the acid-base behavior in that molecules which do not contain hydrogen ions or in non-aqueous medium.

11.1 Lewis acid

According to this theory, an acid is a substance which has capability to accept the non-bonding pair of electrons, called Lewis acid. They are sometimes referred as electron deficient species or electrophile.

11.1.1 Lewis acid: characteristics

- Lewis acid-electron-pair acceptor.
- Lewis acid should have a vacant or empty orbital.
- All cations (Na^+, Cu^{2+}, Fe^{3+}) are Lewis acids because they have capability to accept a pair of electrons but all Lewis acids are not cations.

$$Fe^{3+} + 6H_2O \longrightarrow [Fe(H_2O)_6]^{3+}$$

Metal cation having positive charge and empty orbital	Lone pair present on oxygen

More is the positive charge on the metal, more is the acidic character. Fe^{3+} is good Lewis acid than Na^+.

- The ion, molecule or an atom which has incomplete octet of electrons are also Lewis acids.

For example: BF_3.

Here, you can see that the central atom boron has six electrons in its outermost shell. So, it has ability to accept more electrons due to the presence of an empty orbital and hence, act as Lewis acid.

$$BF_3 + F^- \longrightarrow BF_4^-$$

- The molecule in which the central atom has more than eight electrons (SiF_4, $SiBr_4$), are also considered as Lewis acids.
- The molecule like CO_2, SO_2, etc. are also considered as Lewis acid. These types of molecules form multiple bond between the atoms of different electronegativity. In case of transition metal ions, the metal having more electronegativity makes stronger Lewis acids.
- Electron poor π: system is also considered as Lewis acids, for example, $[CH_2{=}CH]^+$, etc.

11.2 Lewis base

A base is a substance which has capability to donate the electrons, called Lewis base. They are sometimes referred as electron rich species or Nucleophile.

11.2.1 Lewis base: characteristics

1. Lewis base-electron-pair donor

2. All metal anions (F^-, Cl^-, Br^-, I^-) are Lewis base because they have ability to donate the electron but all Lewis bases are not anions.

3. The ion, molecule or an atom which having a lone pair of electrons, are also considered as Lewis base.

4. The electron-rich π system is also considered as Lewis bases, for example, benzene, ethene, etc.

5. The strength of the Lewis base can be increased by increasing the electron density.

Note: When a Lewis acid reacts with Lewis base, then Lewis acid uses its lowest unoccupied molecular orbital (LUMO) and base uses its highest occupied molecular orbital (HUMO) to create a bonded molecular orbital. Actually, Lewis acid and Lewis base both have LUMO and HUMO but HUMO is always considered as base and LUMO is always considered as acid.

$$\text{HOMO} + \text{LUMO} \longrightarrow \text{Bonding MO}$$

11.3 Example of Lewis acid-base

A simplest example of Lewis acid-base is shown by a chemical reaction:

$$Na^+ + Cl^- \longrightarrow Na—Cl$$

In this reaction, chloride ion acts as Lewis base because it has lone pairs of electrons and sodium ion has positive charge, so it acts as Lewis acid.

11.4 Neutralization reaction between Lewis acid and Lewis base

When a Lewis acid reacts with a Lewis base, then a Lewis acid-base reaction occurs in which the molecule which act as Lewis base donate its electron pair into the empty orbital of an acid, forms Lewis acid-base adduct as shown in **Figure 2**. The adduct formed contains a covalent coordinate bond between Lewis acid and Lewis base. The above explanation implies that the Lewis acid is a low electron density centre and Lewis base is a high electron density centre [6].

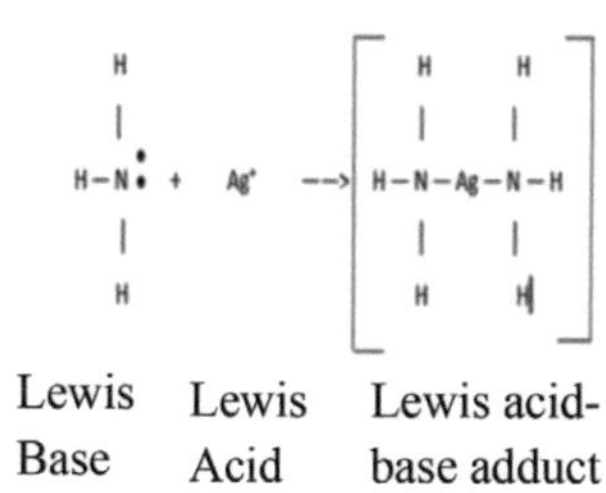

Figure 2.
Acid-base neutralization.

In this reaction, the two ammonia molecules reacts with silver ion. The ammonia has lone pair of electrons, so it has the ability to donate the lone pairs of electrons and acts as Lewis base. The positive charge on silver denotes its electrophilic nature that means it has an ability to accept the pairs of electrons and act as Lewis acid (by Lewis definition).

Here, it can also be noted that when a Lewis acid reacts with a Lewis base, there is no change in the oxidation number of any of these atoms.

11.5 Limitations of Lewis acid-base theory

1. This theory is not able to explain that why all acid-base reactions do not involve the covalent coordination bond.

2. This theory is also unable to explain the behavior of some acids like hydrogen chloride (HCl) and sulfuric acid (H_2SO_4) because they do not form the covalent coordination bond with bases. Hence, they are not considered as Lewis acids.

3. This theory cannot explain the concept that why the formation of coordination bond is a slow process and acid-base reactions is a fast process.

4. This theory cannot explain the concept of relative strength of acids and bases.

5. This theory fails to explain the catalytic activity of some Lewis acids.

12. Relation between Lewis acid-base theory and Arrhenius theory

1. All Arrhenius acids and Bronsted Lowry acids are Lewis acids but reverse is not true.

13. Conclusion

Acids and bases are very important for modern society and in our daily lives. They exist everywhere in our body and in our surroundings. The theory that has been described in this chapter has given us all the basic information of acids and bases. In this chapter, we have discussed all the three basic theory of acid-base chemistry-Arrhenius theory, Bronsted-Lowry theory and Lewis acid-base theory. Acids and bases have vital role in the area of medicine. From this concept, it is now easy to treat the diseases with the improved medicines by complex understanding of acids and bases. For example, If the concentration of hydrogen ion increases in the human blood, acidity increases that results weakness in body. In that condition, the body should keep alkaline by digesting food that produces alkali in the body, to neutralize the acidity.

Definitely, without acids or bases information, our lives would look different to how it looks now. Many products we are using today would have no use without this knowledge.

Acknowledgements

I am eternally grateful and beholden to my family. My mother Mrs. Suman Munjal, Father Mr. Bhim Sain Munjal and sister Mrs. Shweta Java for strengthening me with the opportunities and experiences which enabled me in reaching these heights. The reason behind this success is their selfless encouragement that helped me explore new dimensions in my life.

I especially acknowledge the contribution of my confidence Dr. Himanshu Mathur and brother Mr. Chirag Munjal for his continuous and consistent support, efficient efforts of proof reading my works, kind words of motivation, and providing me with helpful tips. I would like to show my appreciation for sharing his expertise and experiences for organizing required resources.

Conflict of interest

None.

Declaration

I Shikha Munjal undersigned solemnly declare that all the information submitted by me in this chapter is correct, true and valid.

Author details

Shikha Munjal[1] and Aakash Singh[1,2]*

1 Department of Chemistry, School of Basic Sciences, Jaipur National University, Jaipur, India

2 Department of Chemistry, Suditi Global Academy, Manipuri, UP, India

*Address all correspondence to: aakashsingh.ucst@gmail.com

IntechOpen

© 2020 The Author(s). Licensee IntechOpen. This chapter is distributed under the terms of the Creative Commons Attribution License (http://creativecommons.org/licenses/by/3.0), which permits unrestricted use, distribution, and reproduction in any medium, provided the original work is properly cited.

References

[1] Khan Academy. Arrhenius acids and bases [Internet]. Available from: https://www.khanacademy.org/science/chemistry/acids-and-bases-topic/acids-and-bases/a/arrhenius-acids-and-bases

[2] Structure and reactivity. Acidity and basicity [Internet]. Available from: http://www.chem.ucla.edu/~harding/notes/notes_14C_str&react.pdf

[3] Brown TE, Eugene LH, Bursten BE, Murphy C, Woodward P. The Central Science. 11th ed. New York: Prentice-Hall: Chemistry; 2009, ISBN 0136006175

[4] Lumen. Boundless chemistry. Acids and bases [Internet]. Available from: http://courses.lumenlearning.com/boundless-chemistry/chapter/acids-and-bases/

[5] Brewer L. The generalized Lewis acid-base theory: Surprising recent developments. Journal of Chemical Education. 1984;**61**(2):101. DOI: 10.1021/ed061p101

[6] Redefining Knowledge. A hub for gaining knowledge. Theories of acids and bases [Internet]. Available from: https://hemantmore.org.in/science/chemistry/theory-acids-bases/3510/

Chapter 6

Corrosion Wear of Pipelines and Equipment in Complex Stress-Strain State

Vladimir A. Pukhliy

Abstract

An analytical approach is described to determine the resource of structural elements of hydropower, the nuclear industry, etc., under difficult stress conditions when exposed to a corrosive environment with certain parameters (degree of chemical activity, temperature, humidity, flow rate, etc.). The initial-boundary problem for structural elements (bimetallic pipelines, centrifugal pumps) is considered with a decrease in the thickness of the element due to the influence of a corrosive environment. In this case, the effect of a corrosive medium on an element is described by a differential equation with certain initial conditions. Equations describing the stress-strain state of an element are added to this equation. As the first object, the corrosion wear of bimetallic pipelines of nuclear energy is considered. The solution to the problem is to integrate the ordinary differential equation. The criterion for terminating the step-by-step process in time is the condition $\left|\sigma_{ij}\right|_{\min}^{\max} \le [\sigma_T]$, where σ_T is the yield strength of the material of the structural element. As the second object, the corrosion wear of the working blades of centrifugal pumps is considered. The stress-strain state of the blade is described by a system of partial differential equations of the 12th order, to the solution of which the method of integral relations by Dorodnitsyn is applied. At the next stage, the system of ordinary differential equations is integrated by the modified method of successive approximations developed by Professor V.A. Pukhliy. At each stage, the corrosion equation is attached to the solution of this problem. For bimetallic pipelines, a specific example of calculation according to the described algorithm has been implemented.

Keywords: general corrosion, pipelines of hydropower and nuclear energy, corrosion cracking, two-layer shells, initial-boundary value problem, centrifugal pumps, modified method of successive approximation

1. Introduction

In a number of industries, in particular in the chemical industry, axial and radial turbomachines (compressors, superchargers, gas turbine installations, pumps)

operate under aggressive environments, as a result of which rotor blades and disks are subject to corrosion wear.

Pipelines of hydropower are subject to corrosion wear, as a result of which their service life is significantly reduced. The problem of corrosion is especially acute in nuclear energy. Pipelines are made of low-alloy pearlite-grade carbon steel with stainless steel cladding on the inner surface. We also note that all parts and assemblies of the main circulation pumps in contact with the coolant, industrial cooling water, and locking water are made of special steels that are resistant to corrosion.

The coolant of the first circuit of a nuclear reactor is not just pure water, but water with boric acid dissolved in it (H_3BO_3), which contributes to significant corrosion of metal in pipelines. Steam generators of nuclear plants are made of pipes clad with anticorrosive austenitic surfacing.

It should be emphasized that the influence of the stress-strain state on the rate of corrosion and erosion wear also becomes important. For example, a strain of 1% increases the rate of corrosion of silicon iron in a 0.01% solution of sulfuric acid by 53% compared with undeformed metal [1].

Stress corrosion cracking of metals was previously studied. This phenomenon takes place at certain critical (threshold) values of tension determined by the acting stresses and potential energy. Stresses less than critical have an effect on general corrosion without causing cracking.

The impellers of radial and axial turbomachines subjected to corrosion are usually thin-walled plates and shells. The problem of the durability of the elements of the impellers of turbomachines is the problem of the durability of plates and shells of a variable thickness over time, under the influence of an aggressive environment that has certain parameters (degree of chemical activity, temperature, flow rate, etc.), and the stress-strain state.

One of the first works in this direction was an article by Kornishin [2], in which a joint solution of the corrosion equation is considered, which is a linear dependence of the corrosion rate on stress and equations describing the stress-strain state of a shell of variable thickness. The system of joint equations describing the behavior of the shell in a corrosive medium is then solved in finite differences according to a two-layer explicit scheme with a time step.

To date, there are a number of semi-empirical models that approximate corrosion wear taking into account the stress state. In T**able 1**, a number of models used in the calculations are given [3].

Table 1 indicates: h is the depth of the wear layer; t is the time; σ and ε are stress and strain; T is the temperature; k, α, β, and γ are constants; and φ (t) are some functions.

The issues of corrosion wear of centrifugal fan elements were investigated in a number of works by Pukhliy and Semenenko [1, 4, 5].

2. Corrosion cracking of bimetallic pipelines

Bimetallic structures are widely used in modern technology, in particular, in the manufacture of bimetallic elements in nuclear energy. These are, first of all, bimetallic pipelines, bends, etc. These elements are characterized by high strength, heat resistance, and corrosion resistance.

The structural elements of nuclear power units operate under complex loading conditions, in particular, under conditions of exposure to aggressive environments. In this regard, the determination of the time to destruction of structural elements (resource) is the most important in the study of corrosion wear of elements that are in a complex stress-strain state [6–8].

Corrosion equation wear	The law of the effect of stress on corrosion rate
$\frac{dh}{dt} = \varphi(t)(1 + k\sigma)$	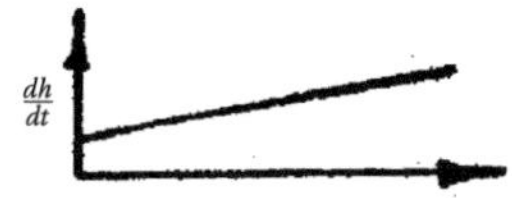
$\frac{dh}{dt} = \varphi(t)c^{\alpha\sigma/RT}$	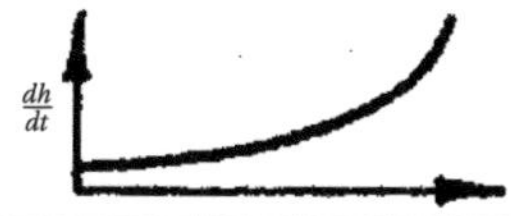
$\frac{dh}{dt} = \alpha + \beta(\sigma - \sigma_0)\ \varepsilon$	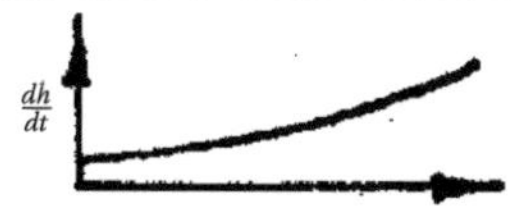
$\frac{dh}{dt} = \alpha - \beta e^{-\gamma\sigma}$	

Table 1.
Corrosion wear models.

In the present paper, an analytical approach to determining the resource of structural elements of nuclear power units based on the theory of bimetallic shells, taking into account the stress-strain state and corrosion wear of the elements, is presented.

Consider a cylindrical bimetallic shell when exposed to a corrosive environment [4]. Such tasks are very important in relation to the design of pipelines of nuclear power plants (corrosion cracking).

The rate of change of thickness at a given point in the shell is taken in the form:

$$\frac{dh}{dt} = F(t, T, \sigma), 0 \le t \le t_k, h > 0. \tag{1}$$

with the initial condition:

$$h(x, y, 0) = h_0(x, y), \tag{2}$$

where x, y are the normal coordinates of the middle surface of the shell; T is the temperature; and σ is the function connecting the rate of change of the shell thickness with the stress state at a surface point.

Note that F is a known function whose form is determined from experiment and t_k is the final point in time.

We study the effect of the stress state on general corrosion under the assumption that the corrosion rate is a linear function of the stress intensity.

The equations of corrosion wear are written as follows:

$$\frac{dh}{dt} = -\alpha\varphi(t)\left(1 + k\sigma_{ij}\right), 0 \le t \le t_k, h > 0. \tag{3}$$

Here σ_{ij} is the stress intensity on the surface of the bimetallic shell; α and k are finite coefficients; and $\varphi(t)$ is a dimensionless function of time. As a rule, in most practical cases $\varphi(t)$ it is a constant or monotonically decreasing function.

It is necessary to add the equations of the theory of bimetallic shells to Eq. (1) or (3). As a result, we obtain an unrelated problem of the theory of shells, in view of which it is possible to apply a finite-difference approximation in time to the solution of Eq. (3).

Thus, the algorithm for solving the initial-boundary-value problem is reduced to the joint solution of Eq. (3) under the initial conditions in Eq. (2) and the system of equations for bimetallic shells, in the general case of variable thickness under the corresponding boundary conditions. Moreover, at each time step from Eq. (3), we obtain numerical values of the thickness of the structural element, which are then used to construct spline functions [9]. Then the system of equations of bimetallic shells is solved, from the solution of which the values of σ_{ij} are determined.

The criterion for terminating the step-by-step process is the following condition:

$$\left| \sigma_{ij}^{\max\atop\min} \right| \leq [\sigma_T],$$

where σ_T is the yield strength of the material of the structural element.

The resource of structural elements of nuclear power units as a whole is determined by the summation of time steps.

We obtain the equilibrium equation of the bimetallic shell on the basis of the Lagrange variational principle:

$$\delta\Pi = \delta\Pi_1 + \delta\Pi_2 = 0. \tag{4}$$

Here $\delta\Pi_1$ is a variation of the potential energy of shell deformation; $\delta\Pi_2$ is a variation of the potential of external forces, equal to the variations of the work of external forces, taken with the opposite sign.

We write an expression for the variation of potential energy:

$$\delta\Pi_1 = \int_{a_1}^{a_2}\int_{b_1}^{b_2} \delta a_0 A_1 A_2 d\alpha_1 d\alpha_2 = \int_{a_1}^{a_2}\int_{b_1}^{b_2} \delta a_0 dS_1 dS_2, \tag{5}$$

where a_0 is the potential energy of deformation of a unit surface of the shell.

Integration extends to the entire surface of the junction (**Figure 1**): from $\alpha_1 = a_1$ to $\alpha_1 = a_2$ and from $\alpha_2 = b_1$ to $\alpha_2 = b_2$.

$$\delta a_0 = \int_0^{\delta_1} \left[\sigma_1^{(1)} \cdot \delta\varepsilon_1^{(1)} + \sigma_2^{(1)}\delta\varepsilon_2^{(1)} + \tau_{12}^{(1)}\delta\gamma_{12}^{(1)}\right]\left(1 - \frac{z}{R_1}\right)\left(1 - \frac{z}{R_2}\right) dz$$
$$+ \int_{-\delta_2}^{0} \left[\sigma_1^{(2)}\delta\varepsilon_1^{(2)} + \sigma_2^{(2)} \cdot \delta\varepsilon_2^{(2)} + \tau_{12}^{(2)}\delta\gamma_{12}^{(2)}\right]\left(1 - \frac{z}{R_1}\right)\left(1 - \frac{z}{R_2}\right) dz$$

In the analysis, the following assumptions are used (**Figure 1**).

The curvilinear coordinate system coincides with the lines of the main curvatures. This coordinate system is a Gaussian coordinate system, it is orthogonal.

The position of the point that does not belong to the junction surface determines the coordinates of the z-distance normal to the point from the junction surface (+ if it is directed along the internal normal to the junction surface).

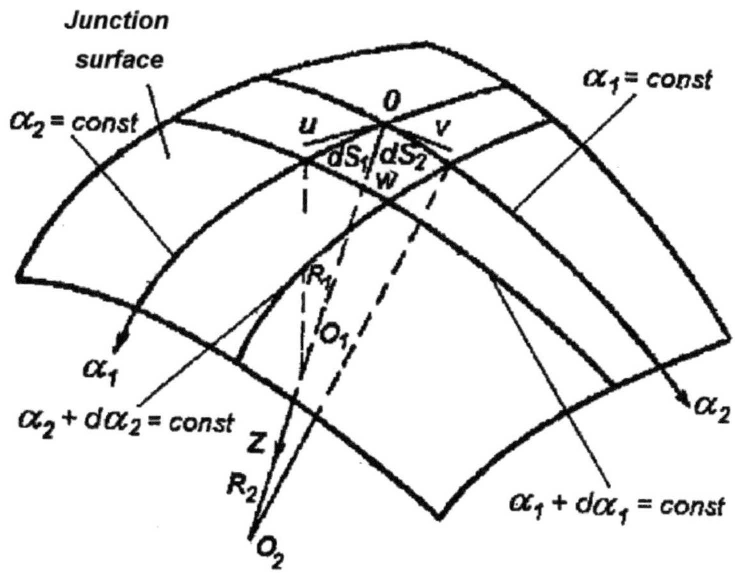

Figure 1.
System of curvilinear coordinates on an undeformed junction surface.

The movement of u and v in are the direction of the tangents to α_1 and α_2 and w in the direction of the normal to the junction surface.

Deformations of the junction surface are determined by relative elongations ε_1 and ε_2 in both the α_1 and α_2 directions, and by a shift γ_{12}—a change in the angle between the tangents to the lines α_1 and α_2 (before deformation $\pi/2$, after $\frac{\pi}{2} - \gamma_{12}$).

Eq. (5) can be represented as follows:

$$\delta\Pi_1 = \int_{a_1}^{a_2}\int_{b_1}^{b_2}(N_1\delta\varepsilon_1 + N_2\delta\varepsilon_2 - M_1\delta\chi_1 - M_2\delta\chi_2 + T\delta\gamma_{12} - H\delta\chi_{12})A_1A_2d\alpha_1d\alpha_2 \quad (6)$$

Denote by:

p_1, p_2, p_3 the projection of external surface forces, referred to the unit of the junction surface, on the direction of the tangents to the lines of curvature α_1 and α_2 and normal to the junction surface;

$N_1^{\alpha_1}, T_1^{\alpha_1}, Q_1^{\alpha_1}, M_1^{\alpha_1}$ the normal, shear, shear forces, and bending moment for the section$\alpha_1 = const$;

$N_2^{\alpha_2}, T_2^{\alpha_2}, Q_2^{\alpha_2}, M_2^{\alpha_2}$ the same for section$\alpha_2 = const$.

Then the variation of the potential of external forces is equal to:

$$\delta\Pi_2 = -\int_{a_1}^{a_2}\int_{b_1}^{b_2}(p_1\delta u + p_2\delta v + p_3\delta w)A_1A_2d\alpha_1d\alpha_2 - \int_{b_1}^{b_2}(N_1^{\alpha_1}\delta u + T_1^{\alpha_1}\delta v - M_1^{\alpha_1}\delta v_1$$
$$+ Q_1^{\alpha_1}\delta w)A_2d\alpha_2 - \int_{a_1}^{a_2}(N_2^{\alpha_2}\delta v + T_2^{\alpha_1}\delta u - M_2^{\alpha_2}\delta v_2 + Q_2^{\alpha_2}\delta w)A_1d\alpha_1. \quad (7)$$

Substituting Eqs. (6) and (7) in Eq. (4) we obtain:

$$\delta\Pi = -\int_{a_1}^{a_2}\int_{b_1}^{b_2}\left\{\left[\frac{\partial N_1 A_2}{\partial \alpha_1} - N_2\frac{\partial A_2}{\partial \alpha_1} + \frac{\partial T_2 A_1}{\partial \alpha_2} + T_1\frac{\partial A_1}{\partial \alpha_2} - \frac{A_1 A_2}{R_1}Q_1 + A_1 A_2 p_1\right]\delta u + \right.$$
$$+\left[\frac{\partial N_2 A_1}{\partial \alpha_2} - N_1\frac{\partial A_1}{\partial \alpha_2} + T_2\frac{\partial A_2}{\partial \alpha_1} - \frac{A_1 A_2}{R_2}Q_2 + A_1 A_2 p_2\right]\delta v +$$
$$\left.+\left[\frac{\partial Q_1 A_2}{\partial \alpha_1} + \frac{\partial Q_2 A_1}{\partial \alpha_2} + A_1 A_2\left(\frac{N_1}{R_1} + \frac{N_2}{R_2}\right) + A_1 A_2 p_3\right]\delta w\right\} d\alpha_1 d\alpha_2 +$$
$$+\int_{b_1}^{b_2}\left\{\left[N_1 - N_1^{\alpha_1}\right]\delta u + \left[T_1 - \frac{H_1}{R_2} - T_1^{\alpha_2}\right]\delta v - \left[M_1 - M_1^{\alpha_1}\right]\delta\left(\frac{1}{A_1}\frac{\partial w}{\partial \alpha_1} + \frac{u}{R_1}\right) + \right.$$
$$\left.+\left[Q_1 + \frac{1}{A_2}\frac{\partial H_1}{\partial \alpha_2} - Q_1^{\alpha_1}\right]\delta w\right\} A_2 d\alpha_2 + \int_{a_1}^{a_2}\left\{\left[T_2 - \frac{H_2}{R_1} - T_2^{\alpha_2}\right]\delta u + \left[N_2 - N_2^{\alpha_2}\right]\delta v - \right.$$
$$\left. - \left[M_2 - M_2^{\alpha_2}\right]\delta\left(\frac{1}{A_2}\frac{\partial w}{\partial \alpha_2} + \frac{v}{R_2}\right) + \left[Q_2 + \frac{1}{A_1}\frac{\partial H_2}{\partial \alpha_1} - Q_2^{\alpha_2}\right]\delta w\right\} - \left\{\left[H_1 + H_2\right]\delta w\right\}_{b_1,a_1} -$$
$$-\left\{\left[H_1 + H_2\right]\delta w\right\}_{b_2,a_2} + \left\{\left[H_1 + H_2\right]\delta w\right\}_{b_2,a_2} + \left\{\left[H_1 + H_2\right]\delta w\right\}_{b_1,a_2}. \tag{8}$$

Here Q_1 and Q_2 are the transverse forces arising in the shell:

$$Q_1 = \frac{1}{A_1 A_2}\left[\frac{\partial M_1 A_2}{\partial \alpha_1} - M_2\frac{\partial A_2}{\partial \alpha_1} + H_1\frac{\partial A_2}{\partial \alpha_2} + \frac{\partial H_2 A_1}{\partial \alpha_2}\right]; \tag{9}$$

$$Q_2 = \frac{1}{A_1 A_2}\left[\frac{\partial M_2 A_1}{\partial \alpha_2} - M_1\frac{\partial A_1}{\partial \alpha_2} + H_2\frac{\partial A_2}{\partial \alpha_1} + \frac{\partial H_1 A_2}{\partial \alpha_1}\right]. \tag{10}$$

The last four terms in Eq. (8) is the work of concentrated forces along the edges of the shell $\alpha_1 = const$, $\alpha_2 = const$.

From Eq. (8), we obtain the equilibrium equation and boundary conditions.

2.1 Equilibrium equations

In the mechanics of a solid deformable body, equilibrium equations can be obtained by making up for the main vector and the main moment of all the forces acting on the element for the infinitely small element extracted from the shell under the influence of external and internal forces (**Figure 2**). Here, the equilibrium equations and boundary conditions are obtained from the variational Lagrange principle in Eq. (4).

Note that in the case of dynamics, it is necessary to apply the variational Hamilton-Ostrogradsky principle.

So, from the first integral of expression in Eq. (8), the first three equations of equilibrium follow:

$$\left.\begin{aligned}
&\frac{1}{A_1 A_2}\left[\frac{\partial N_1 A_2}{\partial \alpha_1} - N_2\frac{\partial A_2}{\partial \alpha_1} + \frac{\partial T_2 A_1}{\partial \alpha_2} + T_1\frac{\partial A_1}{\partial \alpha_2}\right] - \frac{Q_1}{R_1} + p_1 = 0;\\
&\frac{1}{A_1 A_2}\left[\frac{\partial T_1 A_2}{\partial \alpha_1} - N_1\frac{\partial A_1}{\partial \alpha_2} + \frac{\partial N_2 A_1}{\partial \alpha_2} + T_2\frac{\partial A_2}{\partial \alpha_1}\right] - \frac{Q_2}{R_2} + p_2 = 0;\\
&\frac{1}{A_1 A_2}\left[\frac{\partial Q_1 A_2}{\partial \alpha_1} + \frac{\partial Q_2 A_1}{\partial \alpha_2}\right] - \frac{N_1}{R_1} + \frac{N_2}{R_2} + p_3 = 0.
\end{aligned}\right\} \tag{11}$$

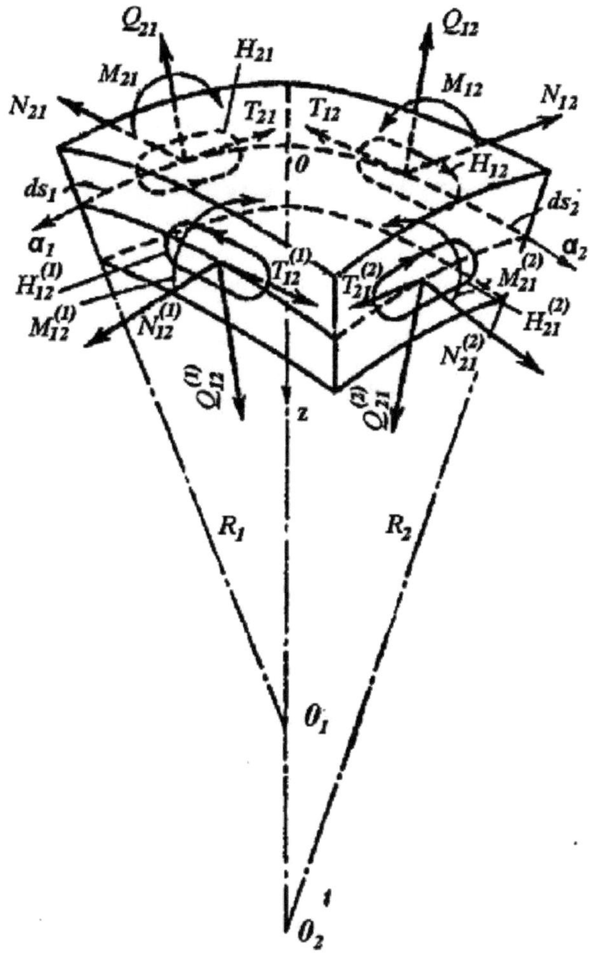

Figure 2.
Internal forces acting on the edge of the element.

From the expressions in Eqs. (9)–(10), two more equations follow:

$$\begin{aligned}
&\frac{1}{A_1A_2}\left[\frac{\partial M_1A_2}{\partial\alpha_1}-M_2\frac{\partial A_2}{\partial\alpha_1}+\frac{\partial H_2A_1}{\partial\alpha_2}+H_1\frac{\partial A_1}{\partial\alpha_2}\right]-Q_1=0;\\
&\frac{1}{A_1A_2}\left[\frac{\partial M_2A_1}{\partial\alpha_2}-M_1\frac{\partial A_1}{\partial\alpha_2}+\frac{\partial H_1A_2}{\partial\alpha_1}+H_2\frac{\partial A_2}{\partial\alpha_1}\right]-Q_2=0.
\end{aligned} \tag{12}$$

The sixth equation is an identity expressing the equality of the moments of all forces acting on the element to zero relative to the axis normal to the surface of the element junction:

$$T_1-T_2-\frac{H_1}{R_1}+\frac{H_2}{R_2}=0.$$

This equation was used to obtain Eq. (11).

2.2 Border conditions

The second and third integrals of expression in Eq. (8) give boundary conditions for the edges $\alpha_1 = const$ and $\alpha_2 = const$, that is, for lines of principal curvatures.

We emphasize that if one of the lines of the main curvatures is closed, then the displacements along this line will be periodic functions.

2.3 Axisymmetric deformation of a cylindrical shell

We introduce the following notation (**Figure 3**): R is the radius of the surface of the junction of a cylindrical bimetallic shell; l is the shell length; δ_1 and δ_2 are the thickness of the inner and outer layers; x is the distance from the left edge of the cylinder to the current section; σ_1 and σ_2 are normal stresses; τ_{13} and is shear stress.

$$M_1 = \int_0^{\delta_1} \sigma_1^{(1)} z dz + \int_{-\delta_2}^{0} \sigma_1^{(2)} z dz; N_1 = \int_0^{\delta_1} \sigma_1^{(1)} dz + \int_{-\delta_2}^{0} \sigma_1^{(2)} dz;$$

$$M_2 = \int_0^{\delta_1} \sigma_2^{(1)} z dz + \int_{-\delta_2}^{0} \sigma_2^{(2)} z dz; N_2 = \int_0^{\delta_1} \sigma_2^{(1)} z dz + \int_{-\delta_2}^{0} \sigma_2^{(2)} z dz;$$

$$Q_1 = \int_0^{\delta_1} \tau_{13}^{(1)} \left(1 - \frac{z}{R}\right) dz + \int_{-\delta_2}^{0} \tau_{13}^{(2)} \left(1 - \frac{z}{R}\right) dz.$$

Relative deformations of a surface located at a distance z from the junction surface (**Figure 4**):

$$\varepsilon_{1z} = \varepsilon_1 - z\chi_1; \ \chi_1 = w''; \ \varepsilon_{2z} = \varepsilon_2 - z\chi_2; \ \chi_2 = \frac{w}{R^2}.$$

Normal stresses according to Hooke's law:

$$\sigma_1^{(1)} = \frac{E_1}{1-\mu_1^2}\left[\varepsilon_1 + \mu_1\varepsilon_2 - z(\chi_1 + \mu_1\chi_2) - (1+\mu_1)\beta_1 \overline{T}\right]; \sigma_2^{(1)} = \frac{E_1}{1-\mu_1^2}\left[\varepsilon_2 + \mu_1\varepsilon_1 - z(\chi_2 + \mu_1\chi_1) - (1+\mu_1)\beta_1 \overline{T}\right]; (0 \le z \le \delta_1)$$

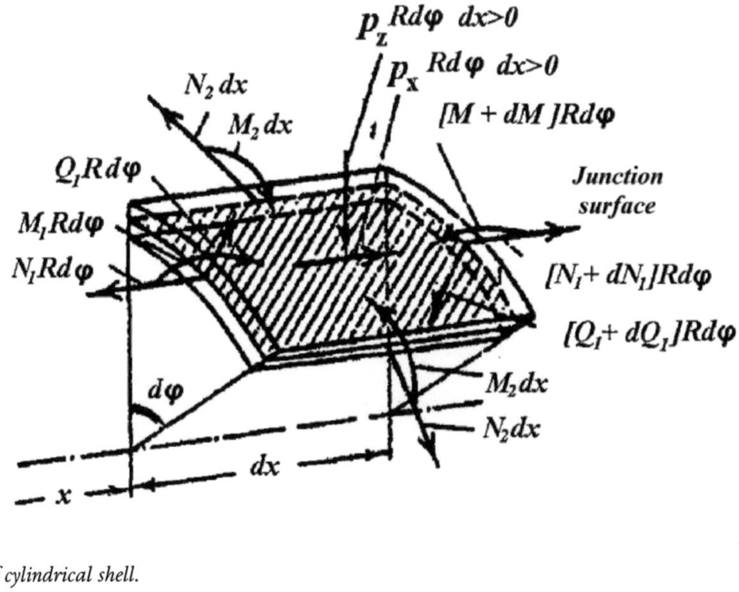

Figure 3.
Element of cylindrical shell.

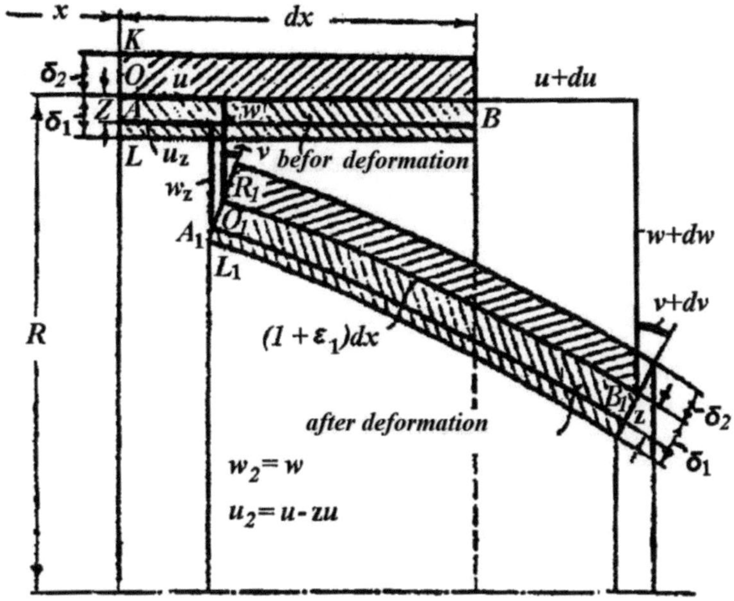

Figure 4.
Deformation of surface junction shell.

$$\sigma_1^{(2)} = \frac{E_2}{1-\mu_2^2}\left[\varepsilon_1 + \mu_2\varepsilon_2 - z(\chi_1 + \mu_2\chi_2) - (1+\mu_2)\beta_2\overline{T}\right]; \sigma_2^{(2)}$$
$$= \frac{E_2}{1-\mu_2^2}\left[\varepsilon_2 + \mu_2\varepsilon_1 - z(\chi_2 + \mu_2\chi_1) - (1+\mu_2)\beta_2\overline{T}\right].(-\delta_2 \le z \le 0)$$

Power factors:

$$M_1 = c_1\varepsilon_1 + c_2\varepsilon_2 - D_1\chi_1 - D_2\chi_2 - g; N_1 = B_1\varepsilon_1 + B_2\varepsilon_2 - c_2\chi_2 - f;$$
$$M_2 = c_2\varepsilon_1 + c_1\varepsilon_2 - D_2\chi_1 - D_1\chi_2 - g; N_2 = B_2\varepsilon_1 + B_1\varepsilon_2 - c_2\chi_1 - c_1\chi_2 - f.$$

Here:

$$B_1 = \frac{E_1\delta_1}{1-\mu_1^2} + \frac{E_2\delta_2}{1-\mu_2^2}; c_1 = \frac{1}{2}\frac{E_1\sigma_1^2}{1-\mu_2^2} - \frac{1}{2}\frac{E_2\delta_2^2}{1-\mu_2^2}; D_1 = \frac{1}{3}\frac{E_1\delta_1^3}{1-\mu_1^2} + \frac{1}{3}\frac{E_2\delta_2^3}{1-\mu_2^2};$$
$$f = \frac{E_1\delta_1 m}{1-\mu_1} + \frac{E_2\delta_2 m}{1-\mu_2}; g = \frac{1}{2}\frac{E_1\delta_1^2 n_1}{1-\mu_1} + \frac{1}{2}\frac{E_2\delta_2^2 n_2}{1-\mu_2};$$
$$m_1 = \frac{1}{\delta_1}\int_0^{\delta_1}\beta_1 t dz; m_2 = \frac{1}{\delta_2}\int_{-\delta_2}^{0}\beta_2 t dz; n_1 = \frac{2}{\delta_1^2}\int_0^{\delta_1}\beta_1 t z dz; n_2 = \frac{2}{\delta_2^2}\int_{-\delta_2}^{0}\beta_2 t z dz.$$

Equations of an infinitesimal element:

$$\left.\begin{aligned} &[N_1(1+w')]' + p_x = 0; \\ &Q_1' + \frac{M_2}{R^2} + [N_1 w']' + \frac{N_2}{R} + p_z = 0; \\ &M_1' - Q_1 = 0. \end{aligned}\right\}$$

The problem of axisymmetric deformation of an elastic bimetallic cylindrical shell for any relations between thicknesses, different mechanical characteristics of the material of the layers, and arbitrary heating along the thickness and axial direction is described by the equation:

$$w^{IV} + 2aw'' + b^2 w = \theta(x). \tag{13}$$

$$a = \frac{1}{R}\frac{c_2 B_1 - c_1 B_2 - \frac{1}{2} B_1 R N_1}{D_1 B_1 - c_1^2}; b^2 = \frac{1}{R^2}\frac{B_1^2 - B_2^2}{D_1 B_1 - c_1^2};$$

$$\theta(x) = \frac{B_1}{D_1 B_1 - c_1^2}\left\{\frac{c_1}{B_1}(N_1 + f)'' + \frac{1}{R}\frac{B_2}{B_1}\left[N_1 + \left(1 - \frac{B_1}{B_2}\right)f\right] - g'' - p_z\right\}.$$

Consider the case:

$$c_1 = c_2 = 0; \mu_1 = \mu_2 = \mu; E_1 \delta_1^2 = E_2 \delta_2^2.$$

Then the force factors are written as follows:

$$M_1 = -D[w'' + (1+\mu)n]; N_1 = B\left[\varepsilon_1 - \mu\frac{w}{R} - (1+\mu)m\right];$$

$$M_2 = -D[\mu w'' + (1+\mu)n]; N_2 = \mu N_1 - \frac{B}{R}(1-\mu^2)(w + mR).$$

Here

$$B = \frac{\delta\sqrt{E_1 E_2}}{1-\mu^2}; D = \frac{1}{3}\frac{E_1 E_2 \delta^3}{(1-\mu^2)\left(\sqrt{E_1} + \sqrt{E_2}\right)^2};$$

$$m = \frac{m_1 + m_2\sqrt{E_2/E_1}}{1 + \sqrt{E_2/E_1}}; n = \frac{3}{2}\frac{n_1 + n_2}{\delta};$$

$$k^4 = \sqrt[4]{\frac{(1-\mu^2)B}{4DR^2}} = \frac{\sqrt[4]{3(1-\mu^2)}}{\sqrt{2R\delta}}\frac{\sqrt{\sqrt{E_1} + \sqrt{E_2}}}{\sqrt[3]{E_1 E_2}}.$$

In this case, instead of Eq. (13), we obtain the following equation:

$$w^{IV} - \frac{RN_1}{D}w'' + 4k^4 w = \frac{4k^4 B}{(1-\mu^2)B}(\mu N_1 + p_z R) - 4k^4 Rm - (1+\mu)n''. \tag{14}$$

The boundary conditions for bimetallic shells coincide with similar conditions for homogeneous shells.

So, for a hard-pressed edge we get:

$$w = \frac{\partial w}{\partial n} = 0$$

For the free edge we have:

$$M_1 = Q_1 = 0$$

2.4 Infinitely long cylindrical shell (piping)

We write the solution of the homogeneous Eq. (13) in the following form:

$$w = k_1 ch\alpha x \cos\beta x + k_2 sh\alpha x \sin\beta x + k_3 sh\alpha x \cos\beta x + k_4 ch\alpha x \sin\beta x, \qquad (15)$$

k_1, ... k_4 are arbitrary constants; $\alpha = \sqrt{\frac{b-a}{2}}$; $\beta = \sqrt{\frac{b+a}{2}}$.

For an infinitely long shell in solution in Eq. (15),$sh\alpha x$ and$ch(\alpha x)$ we express it in terms of exponential functions. As x → ∞, terms containing exp. αx → ∞ and integration constants → 0.

Then:

$$w = k_1 A_1(\alpha x, \beta x) + T_2 A_2(\alpha x, \beta x) + w_0.$$

Here

$$A_1 = A_1(\alpha x, \beta x) = e^{-\alpha x}\sin\beta x; A_2 = A_2(\alpha x, \beta x) = e^{-\alpha x}\cos\beta x.$$

Example 1. Consider a bimetallic cylindrical shell under the influence of internal pressure q and a corrosive medium (**Figure 5**).

In this case, the stress intensity is constant for all points of the shell and is equal to [5]:

$$\sigma_i = \frac{qR_a^2}{R_c^2 - R_a^2} - \frac{Q}{R_c^2 - R_b^2}; Q = q\frac{\left(\frac{1}{E_1} - \frac{1}{E_2}\right)\frac{R_b^2 - R_a^2}{R_c - R_a^2}}{E_1^{-1}\left(R_c^2 - R_b^2\right)^{-1} + E_2^{-1}\left(R_b^2 - R_a^2\right)^{-1}}; \qquad (16)$$

Given Eq. (16), Eq. (3) takes the form:

$$\frac{dh}{dt} = -\alpha\varphi(t)\left(1 + \frac{kA}{h}\right); \qquad (17)$$

$$A = q(R_c - R_a)\left\{\frac{R_a^2}{R_c^2 - R_a^2} - \frac{\left(\frac{1}{E_1} - \frac{1}{E_2}\right)\left(\frac{R_b^2 - R_a^2}{R_c^2 - R_a^2}\right)}{\frac{1}{E_1\left(R_c^2 - R_b^2\right)} + \frac{1}{E_2\left(R_b^2 - R_a^2\right)}} \cdot \frac{1}{R_c^2 - R_b^2}\right\}$$

with the initial condition:

$$h(0) = h_0, h_0 = R_c - R_a. \qquad (18)$$

The initial thickness h_0 is taken constant for all points of the two-layer shell.

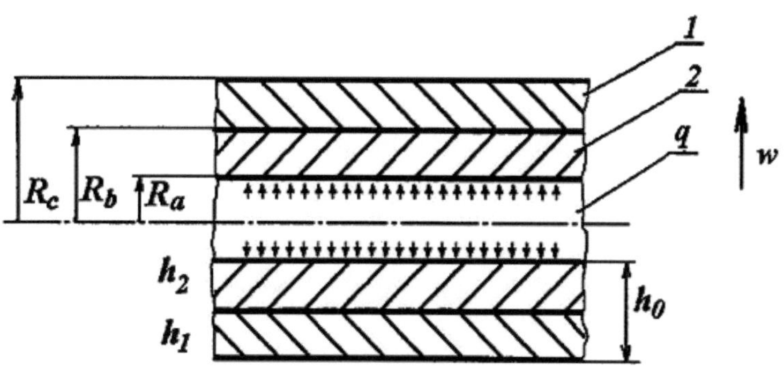

Figure 5.
Geometry and loads acting on an element of two-layer shell: 1—outer layer; 2—inner layer; q_a—internal pressure.

Eq. (17) with the initial condition Eq. (18) is integrated in quadratures:

$$h = h_0 - kP\ln\frac{h_0 kA}{h + kA} - \alpha\int_0^{\tau}\varphi(t)dt, h > 0. \tag{19}$$

In case $\varphi \equiv 1$, expression in Eq. (19) takes the form:

$$h = h_0 - \alpha t - kP\ln\frac{h_0 + kA}{h + kA}, h > 0. \tag{20}$$

We note that Eq. (20) differs from linear equation $h = h_0 - \alpha t$ by the presence of an additional term that takes into account the effect of the stress state of the two-layer shell on the corrosion rate.

Let us determine the durability of a steel two-layer cylindrical shell with R_c = 80.0 cm; $h_0 = R_c - R_a = 0.8$ cm; $h_1 = R_c - R_b = 0.5$ cm; $h_2 = R_b - R_a = 0,3$cm; q = 10 kg/cm^2; and $[\sigma_T]$ = 2200 kg/cm^2; $E_1 = E_2 = 2 \cdot 10^6$ kg/cm^2.

Corrosion rate, $\left|\frac{dh}{dt}\right|_{\sigma_i} = \alpha = 0.03$ cm/year [2]; $\left|\frac{dh}{dt}\right|_{\sigma_i=[\sigma_T]} = \alpha_1 = 0.05$cm/year [2], where $[\sigma_T]$ is the allowable value of stress intensity at the end of the service life.

Given Eq. (3) we get:

$$k = \frac{\alpha_1/\alpha - 1}{[\sigma_T]}.$$

The final value of the thickness of the shell h_T is found from the conditions of achievement $\sigma_{ij} = [\sigma_T]$:

$$h_T = \frac{A}{[\sigma_T]} = 0.315 \text{ cm}.$$

Substituting the values of the coefficients in Eq. (3), we find the durability T = 11.6 years. In conventional calculations, the durability is calculated by the formula: $T = \frac{h_0 - h_T}{\alpha_1}$, where T = 9.7 years.

In conclusion, it should be noted that in the general case it is necessary to solve the unrelated problem of the theory of shells [10, 11], when at each step of integration over time it is necessary to solve the problem of the stress-strain state of a bimetallic shell by a variable thickness, for which it is necessary to use methods of integrating partial equations derivatives [12, 13].

3. The durability of the rotor blades of centrifugal pumps when exposed to corrosion wear

The impellers of centrifugal pumps subjected to corrosion are usually thin-walled plates and shells. The problem of the durability of the elements of the impellers of centrifugal pumps is the problem of the durability of the plates and shells of a variable thickness over time, under the influence of an aggressive environment with certain parameters (degree of chemical activity, temperature, flow rate, etc.), and the stress-strain state.

Figure 6 shows the layout of a centrifugal pump. Consider the durability of the working blades of centrifugal pumps, which are a trapezoidal shell of variable stiffness (**Figure** 7). The blade is subject to the combined action of centrifugal load and corrosion wear.

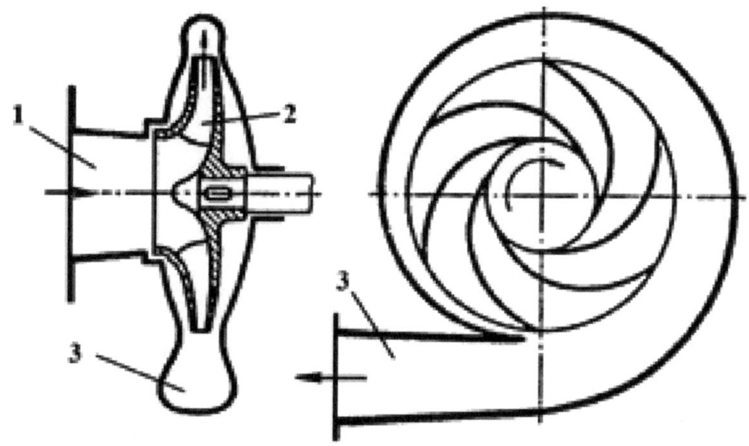

Figure 6.
Scheme of centrifugal pump: 1—confuser; 2—impeller; 3—diffuser.

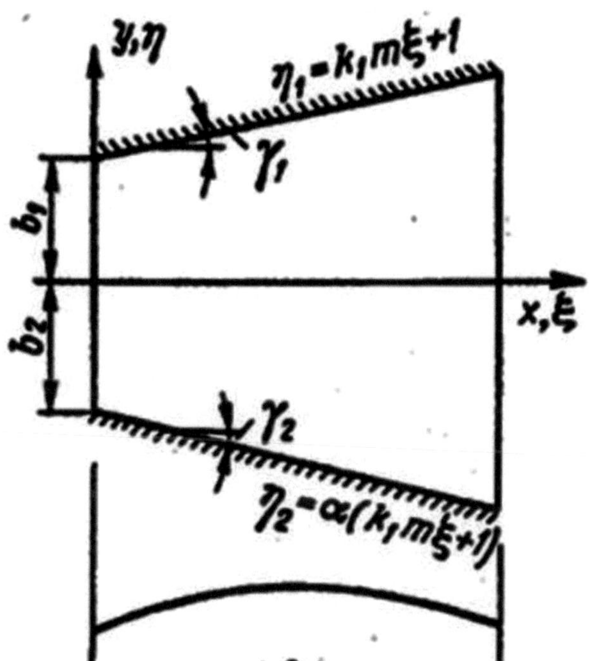

Figure 7.
Geometry and coordinate system of centrifugal pump blade.

The rate of change of thickness at a given point of the blade is taken in the form of a functional relationship:

$$\frac{dh}{dt} = F(t, T, \sigma) \tag{21}$$

with the initial condition:

$$h(x, y, 0) = h_0(x, y), \tag{22}$$

where x, y are the coordinates of the middle surface of the scapula; T is the temperature; and σ is the function connecting the rate of change of thickness with the stress state at a surface point. Function F should be determined from experiment.

Assuming that the rate of change in corrosion wear is a linear function of stress intensity, we write Eq. (21) in the form:

$$\frac{dh}{dt} = \varphi(t)\left(1 + k\sigma_{ij}\right). \tag{23}$$

Eq. (23) must be supplemented with shell theory equations of variable thickness.

Omitting the intermediate calculations, we present a system of partial differential equations of the type Margherra [14] with respect to the normal deflection w and the stress function F of the eighth order, describing the stress state of the blade of variable thickness, taking into account the temperature effect:

$$D\nabla^4 w + 2\frac{\partial D}{\partial x}\frac{\partial}{\partial x}\nabla^2 w + 2\frac{\partial D}{\partial y}\frac{\partial}{\partial y}\nabla^2 w + \nabla^2 D\nabla^2 w - (1-\nu)L(D,W) - h\nabla_k^2 F$$

$$= q - \nabla^2 M_T; \ \frac{1}{B}\nabla^4 F + 2\frac{\partial}{\partial x}\left(\frac{1}{B}\right)\frac{\partial}{\partial x}\nabla^2 F + 2\frac{\partial}{\partial y}\left(\frac{1}{B}\right)\frac{\partial}{\partial y}\nabla^2 F + \nabla^2\left(\frac{1}{B}\right)\nabla^2 F$$

$$- (1+\nu)L\left(F, \frac{1}{B}\right) + \frac{(1-\nu^2)}{h}\nabla_k^2 w$$

$$= -(1-\nu)\frac{1}{h}\nabla^2\left(\frac{N_T}{B}\right). \tag{24}$$

Here T = T(x, y, z) is the temperature field of a general form; α is the coefficient of linear expansion of the material of the blade. ∇^2 and ∇^4 harmonic and biharmonic operators; and ∇_k^2 metaharmonic operator:

$\nabla_k^2 = k_{11}\frac{\partial^2}{\partial x^2} + k_{22}\frac{\partial^2}{\partial y^2}$; L(D, w) and $L\left(\text{F}, \frac{1}{\text{B}}\right)$ are second-order linear differential operators.

Power factors due to temperature exposure are recorded as:

$$N_T = \frac{\alpha E}{1-\nu}\int_{-h/2}^{h/2} T(x,y,z)\,\text{dz};\, M_T = \frac{\alpha E}{1-\nu}\int_{-h/2}^{h/2} T(x,y,z)\,\text{zdz}. \tag{25}$$

We introduce the dimensionless coordinate system:

$$\xi = x/l; \ \eta = y/b_1; \ m = l/b_1 \tag{26}$$

and dimensionless unknown functions

$$\overline{w} = w/h_0; \ \overline{F} = \frac{F}{E^* h_0^2}. \tag{27}$$

Here h_0 is the characteristic thickness of the scapula and E^* is the modulus of elasticity of the material of the blade at a temperature of T = 20°C.

The boundary conditions at the edges of the blade adjacent to the disks $\eta_1 = k_1 m\xi + 1$ and $\eta_2 = \alpha(k_1 m\xi + 1)$ are considered as follows:

$$\overline{w} = \overline{w}_{,n} = \overline{F} = \overline{F}_{,n} = 0. \tag{28}$$

The boundary conditions at the inlet and outlet edges of the blade correspond to the free edge. In this case we have:

$$N_{\xi} = N_{\xi\eta} = M_{\xi} = 0; Q_{\xi} - \frac{\partial M_{\xi\eta}}{\partial \eta} = 0. \tag{29}$$

The important issue is to specify the function of changing the blade thickness h = h (x, y) in the process of erosion-corrosion wear.

In our studies, the function of changing the blade thickness was set in the form of cubic splines [9].

In the general case, the blade thickness can be represented as two-dimensional spline interpolations:

$$h(\xi_i, \eta_i) = h_0 \sum_{\alpha=0}^{3} \sum_{\beta=0}^{3} A_{\alpha\beta}^{(ij)} (\xi - \xi_i)^{\alpha} \left(\eta - \eta_j\right)^{\beta} \ (\mathrm{i} = 0, 1, \dots, \mathrm{n}; \mathrm{j} = 0, 1, \dots, \mathrm{m}). \tag{30}$$

This function on each element of the surface of the scapula

$$\left[\xi_i, \ \xi_{\mathrm{i}+1}; \ \eta_{\mathrm{j}}, \eta_{\mathrm{j}+1}\right]$$

is a bicubic polynomial, continuous, and has continuous partial derivatives up to and including $\frac{\partial^4 h(\xi,\eta)}{\partial\xi^2 \partial\eta^2}$, that is $h(\xi, \eta) \in c^{2,2}$.

We represent the system of Eq. (24) in a dimensionless form:

$$\left.\begin{aligned} L_{11}(\overline{w}) + L_{12}\left(\overline{F}\right) = L_1; \\ \mathrm{L}_{21}(\overline{w}) + L_{22}\left(\overline{F}\right) = L_2, \end{aligned}\right\} \tag{31}$$

where $L_{ik}(i, k = 1, 2)$ are dimensionless differential operators of the theory of shells, referred to the lines of curvature of the surface; L_m $(\mathrm{m} = 1, 2)$ are components of a given surface and temperature load.

The analytical solution of the system of Eq. (31) with boundary conditions in Eq. (28) is based on the application of the method of integral relations by Dorodnitsyn [13].

In accordance with the method, we write the initial system of Eq. (31) in divergent form:

$$\frac{\partial \overline{X}}{\partial \xi} + \frac{\partial \overline{Y}}{\partial \eta} + \overline{L} = 0, \tag{32}$$

where $\overline{X} = \{X_i\} = \left\{\overline{w}, \mathrm{z}_1, z_2, z_3, \overline{F}, z_4, z_5, z_6\right\}$;

$$\overline{Y} = B_0 \overline{X} + B_1 \frac{\partial \overline{X}}{\partial \eta} + B_2 \frac{\partial^2 \overline{X}}{\partial \eta^2} + B_3 \frac{\partial^3 \overline{X}}{\partial \eta^3}; \overline{L} = B\overline{X} + \overline{b}.$$

Through $z_1, \dots, z_6$ marked:

$$z_1 = \overline{w}_{,1}; z_2 = \overline{w}_{,11}; z_3 = \overline{w}_{,111}; z_4 = \overline{F}_{,1}; z_5 = \overline{F}_{,11}; z_6 = \overline{F}_{,111}.$$

In Eq. (32) $B_r = \{b_{mn,s}\}$ and $B = \{b_{mn}\}$;(s = 0, 1, 2, 3; m, n = 1, 2, ..., 8) are transformation matrices.

Following the method of integral relations of Dorodnitsyn [13], we look for a solution to the system of Eq. (32) in the form of an expansion:

$$X_i(\xi,\eta) = \begin{cases} \sum_{j=1}^{n} X_{ij}(\xi)P_j(\xi,\eta) & (\mathrm{i}=3,6,7,8) \\ \sum_{j=1}^{n} X_{ij}(\xi)P_{j,2}(\xi,\eta) & (\mathrm{i}=1,2) \\ \sum_{j=1}^{n} X_{ij}(\xi)P_{j,22}(\xi,\eta) & (\mathrm{i}=4,5) \end{cases} \quad (33)$$

As approximating and weighting functions, we choose the Jacobi system of orthogonal polynomials [15, 16] and their derivatives:

$$P_j(\xi,\eta) = P_1(\xi,\eta)\sum_{j=1}^{n}\left[\eta - \frac{(1+\alpha)r}{2}\right]^{j-1};$$

$$P_1(\xi,\eta) = \eta^4 - 2(1+\alpha)r\eta^3 + \left(1+4\alpha+\alpha^2\right)r^2\eta^2 - 2\alpha(1+\alpha)r^3\eta + \alpha^2 r^4.$$

Here$r = 1 + k_1 m\xi$ is the equation of the inclined side of the scapula.

Note that polynomials $P_j(\xi,\eta)$ are orthogonal on the interval $[\eta_1, \eta_2]$ and forming a system of linearly independent functions, they satisfy the boundary conditions at the oblique edges of the blade in Eq. (28).

We also emphasize that their derivatives $\mathrm{P}_{j,2}(\xi,\eta)$ and $P_{j,22}(\xi,\eta)$ also have the property of orthogonality.

Restricting ourselves to the two-term approximation and also choosing power polynomials $P_j(\xi,\eta)$ and their derivatives as weight functions, after applying the procedure of the method of integral relations to the original system of Eq. (32), we obtain a system of ordinary differential equations of order 8n with variable coefficients:

$$\frac{dX_m}{d\xi} = \sum_{\nu=1}^{s} B_{\nu,m}X_\nu + f_m, \quad \mathrm{m} = 1, 2, \ldots, \mathrm{s} \quad (34)$$

It should be noted that in the general case there is no exact solution of such equations in mathematics, with the exception of individual special cases, for example, the Bessel equation.

Here, the modified method of successive approximations developed by Professor Pukhliy and published by him in the Academic Press [12, 17] is applied to the solution of the boundary-value problem.

Later, the method was extended to the solution of initial-boundary value problems [4], and to accelerate the convergence of the solution, the method of telescopic shift of the power series of K. Lanczos [18] was used. For this, we used the possibility of representing any power series in terms of shifted Chebyshev polynomials [15, 16]. For the first time, such an approach was presented in the works of V.A. Pukhliy [19, 20].

In accordance with the method, variable coefficients $B_{\nu,m}$ and free terms f_m can be represented through shifted Chebyshev polynomials:

$$f_m = \sum_{r=0}^{q} f_{m,r}(d_r \cdot r!)^{-1}\sum_{k=0}^{r} a_k T_k^*(\xi), \quad B_{\nu,m} = \sum_{r=0}^{q} b_{\nu,m,r} d_r^{-1}\sum_{k=0}^{r} a_k T_k^*(\xi) \quad (35)$$

Here q is the degree of the interpolation polynomial and a_k are coefficients of expansion of ξ^r in a series of Chebyshev polynomials $T_k^*(\xi)$. In expressions of Eq. (35), $d_r = 1$ for $r = 0$ and $d_r = 2^{2r-1}$ for the remaining r.

The general solution of the system of Eq. (34) has the form [10, 19, 20]:

$$X_m = \sum_{\mu=1}^{s} C_\mu \left[d_0^{-1} a_0 T_0^*(\xi)\,\delta + \sum_{n=1}^{\infty} X_{m,\mu,n} \right] + \sum_{j=0}^{q} t_{m,j,0} \left[d_{j+1}(j+1)! \right]^{-1} \sum_{k=0}^{j+1} a_k T_k^*(\xi) + \sum_{n=2}^{\infty} X_{m,n}, \tag{36}$$

where $t_{m,j,0} = f_{m,r}$ for $j = r$; μ is the number of the fundamental function; and C_μ are integration constants. In solution in Eq. (36), $\delta = 1$ if $m = \mu$ and $\delta = 0$ for the remaining μ.

The first approximation $X_{m,\mu,1}$ is obtained by substituting the zeroth approximation: $d_0^{-1} a_0 T_0^*(\xi)\,\delta$ into the right-hand side of system $\frac{dX_m}{d\xi} = \sum_{\nu=1}^{s} B_{\nu,m} X_\nu$.

Subsequent approximations are carried out according to the formulas:

$$X_{m,\mu,n} = \sum_{j=1}^{\beta} t_{m,\mu,n,j} \left[d_{n+j-1}(n+j-1)! \right]^{-1} \sum_{k=0}^{n+j-1} a_k T_k^*(\xi);$$
$$X_{m,n} = \sum_{j=1}^{\beta} t_{m,n,j} \left[d_{n+j-1}(n+j-1)! \right]^{-1} \sum_{k=0}^{n+j-1} a_k T_k^*(\xi), \quad \text{where } \beta = n(q+3)\,)-2 \tag{37}$$

The systems of fundamental functions in Eq. (37) are uniformly converging series, and the coefficients $t_{m,\mu,n,j}$ and $t_{m,n,j}$ are determined through the coefficients of the previous approximation using recurrence formulas:

$$t_{m,\mu,n,j} = \sum_{\nu=1}^{s} \sum_{r=0}^{q} b_{\nu,m,r} t_{\nu,\mu,n-1,j-r} (n+j-1)^{-1} \prod_{\gamma=0}^{r} (n+j-1-\gamma),$$
$$t_{m,n,j} = \sum_{\nu=1}^{s} \sum_{r=0}^{q} b_{\nu,m,r} t_{\nu,n-1,j-r} (n+j)^{-1} \prod_{\gamma=0}^{r} (n+j-\gamma).$$

The constants C_μ included in the general solution in Eq. (36) are found from the boundary conditions along the inlet and outlet edges of the blade in Eq. (29).

Thus, the problem reduces to the joint solution of Eq. (23) and the system of Eq. (24) under initial conditions in Eq. (22) and boundary conditions in Eqs. (28) and (29). Moreover, at each time step, from Eq. (23) we obtain the numerical values of the thickness, which are used to construct spline functions in Eq. (30). Then the system of Eq. (24) is solved, from the solution of which the values σ_{ij} are obtained.

The criterion for terminating the step-by-step process is the condition:

$$\sigma_{ij} \le [\sigma_T],$$

where σ_T is the yield strength of the material.

The durability of the impeller element of centrifugal pumps is obtained by summing the steps in time.

4. Conclusions

The theory of corrosion wear of structural elements of hydropower and nuclear energy in the form of plates and shells is developed taking into account the stress state and corrosion wear.

Numerous factors affecting the speed of the corrosion wear process (degree of aggressiveness of the media, temperature, humidity, etc.) are taken into account in a generalized way by drawing up a differential equation for the rate of change of the thickness of the impeller element.

The criterion for the ultimate state of structural elements is the achievement by the structural element of the yield strength of the material σ_T.

An algorithm has been developed for solving the problem of corrosion wear of bimetallic pipelines of nuclear energy, taking into account the stress-strain state of the elements.

An algorithm has been developed for the analytical solution of the problem of corrosion wear of rotor blades of centrifugal pumps based on a combination of the method of integral relations and the modified method of successive approximations in displaced Chebyshev polynomials.

Acknowledgements

Research has been funded by RFBR and City of Sevastopol under Research Project №18-48-920002.

Author details

Vladimir A. Pukhliy
Sevastopol State University, Sevastopol, Russia

*Address all correspondence to: pu1611@rambler.ru

IntechOpen

© 2020 The Author(s). Licensee IntechOpen. This chapter is distributed under the terms of the Creative Commons Attribution License (http://creativecommons.org/licenses/by/3.0), which permits unrestricted use, distribution, and reproduction in any medium, provided the original work is properly cited.

References

[1] Pukhliy VA, Semenenko VI, Zhuzhman VY. New in anti-corrosion protection of elements of industrial ventilation systems. In: Theses of the report on Vses. Scientific and Technical Conf. "Corrosion protection and environmental protection." M.; 1980

[2] Kornishin MS, Karpunin VG, Kleschev SI. To the calculation of plates and shells, taking into account general corrosion. In: Proceedings of X All. Conference on the Theory of Shells and Plates. Vol. 2. Tbilisi: Metsnireba Publishing House; 1975. pp. 71-78

[3] Ovchinnikov IG, Sabitov HA. A comparative study of the extrapolation capabilities of some models of corrosion wear on the example of the calculation of a cylindrical shell. Proceedings of Universities. Construction and Architecture. 1986;**1**:42-45

[4] Pukhliy VA, Nusuev VN, Semenenko VI. Corrosion-resistant fans with a polymer protective coating. In: Improving the Efficiency of Fan Installations. M.; 1982. pp. 111-116

[5] Pukhliy VA. To the theory of erosion-corrosion wear. In: Questions of the Aerodynamics of Textile Equipment. M.: VNIIL Tekmash; 1988. pp. 122-135

[6] Matveenko AF, Baldin AV, Grigoriev AP. Stress corrosion cracking of steels of main gas pipelines. I. Emergency destruction. Physics of Metals and Metallurgy. 1998;**86**(2): 24-31

[7] Matvienko AF, Baldin AV, Grigoriev AP. Stress corrosion cracking of steels of main gas pipelines. II. On the relationship of mechanical properties and resistance of SCC. Metal Physics Metallurgy. 1998;**86**(2):147-155

[8] Rabotnov YN. About a possible mechanism of metal destruction in a corrosive environment. In: Selected Works "Problems of the Mechanics of a Deformable Solid". M.: Publishing House "Science"; 1991. pp. 75-80

[9] Alberg J, Nielson E, Walsh J. Theory of Splines and its Applications. M.: Mir Publishing House; 1972. pp. 318

[10] Pukhliy VA. Solving initial boundary value problems of mathematical physics by the modified method of successive approximations. Review of Applied and Industrial Mathematics. 2017;**24**(1):37-48

[11] Pukhliy VA, Miroshnichenko ST, Zhuravlev AA. Thermoelasticity of the blades of centrifugal pumps of nuclear power under the influence of high-temperature and radiation fields. In: Proceedings of the 12th All-Russian Congress on Fundamental Problems of Theoretical and Applied Mechanics; Vol. 3. Ufa: RAS; 2019. pp. 1203-1205

[12] Pukhliy VA. A method for the analytical solution of two-dimensional boundary value problems for systems of elliptic equations. USSR Computational Mathematics and Mathematical Physics. 1978;**18**(5):1275-1282

[13] Dorodnitsyn AA. On a method for solving the equations of a laminar boundary layer. Journal. Adj. Mate. and Tech. Physics. 1960;**3**:111-118

[14] Marguerre K. Zur theorie der gekrümmten Platte großer Formänderung. In: Jahrbuch 1939 der deutschen Luftfahrforschung. Vol. i. 1939. pp. 413-418

[15] Courant R, Hilbert D. Methods of Mathematical Physics. Vol. I. M.: Gostekhizdat; 1953

[16] Luke VL. Mathematical Functions and Their Approximations. New York,

London: Academic Press. Inc.; 1975. pp. 608

[17] Pukhliy VA. About one approach to solving boundary value problems of mathematical physics. Differential Equations. 1979;**15**(11):2039-2043

[18] Lantsosh K. Practical Methods of Applied Analysis. M.: Fizmatgiz; 1961. pp. 524

[19] Pukhliy VA. The solution to the problem of bending oblique in terms of a cylindrical three-layer panel by the modified method of successive approximations. Applied Mechanics. 1986;**22**(10):62-67

[20] Pukhliy VA. An analytical method for solving boundary value problems of shell theory. In: Proceedings of the XIII All-Union. Conference on the Theory of Plates and Shells. Part IV-Tallinn. TPI Publishing House; 1983. pp. 101-107

Chapter 7

Spectroscopy in Oilfield Corrosion Monitoring and Inhibition

Ekemini Ituen, Lin Yuanhua, Ambrish Singh and Onyewuchi Akaranta

Abstract

Interaction of surfaces of metals and alloys with electromagnetic radiation produces several interesting phenomena, including electronic transitions, molecular rotation and vibration within bonds, polarisation, or photochemical reactions. Spectroscopy is an essential tool that provides some structural information about these interactions. In corrosion, spectroscopic techniques are often employed for mechanistic determinations, especially in the presence of corrosion inhibitors. In this chapter, we have examined some spectroscopic methods that are useful for corrosion monitoring and corrosion inhibition. More emphasis is placed on sample preparation, output parameters, interpretation of results and possible deductions/predictions, which could be made from obtained results than on the underlying principles and mode of equipment operation. Attempts are also made to critically examine some literature, hence readers (early career researchers) and experts in the field will find this chapter very resourceful and a ready reference material.

Keywords: corrosion products, EDS, EIS, FTIR, UV-Vis, XRD

1. Introduction

Once existing wells begin to deplete, it becomes essential to use some principles of chemistry and materials science to maintain production through operations, such as well acidising, secondary and enhanced oil recovery. A common field technique for achieving this is to force a fluid (mixtures which could include acid, water and polymers) through the well bore at high pressure. This fluid dissolves formation rocks, enlarges existing flow channels and opens new ones by etching to enhance the flow of hydrocarbons. During this procedure, corrosion of steel structural materials often occurs. Such materials include line pipes, casings and tubings. After recovery, the spent acid could also cause corrosion of storage facilities. Problems associated with such corrosion could include rupturing of materials, spills, failure, loss of hydrocarbons and loss of materials' integrity and flow challenges. The cost of maintaining/cleaning spills is often high, hence most industries would find methods to prevent or retard corrosion. Also, the downtime due to the shutdown of plant for maintenance and damage to employees or company integrity is undesirable. The use of corrosion inhibitors (CIs) has been considered the simplest and cost-effective means of mitigating oilfield corrosion. Corrosion inhibitors (CIs) are chemical substances which are added in small amounts to the corroding fluid in order to retard the speed of corrosive attack on the steel surface which it contacts [1].

Corrosion inhibitors offer surface protection by adsorption of their active functionalities on the substrate (steel) surface to 'blanket' it from the aggressive medium. The effectiveness or efficiency of the inhibitor varies as its chemical identity, composition, concentration and operational temperature and pressure varies [2, 3]. To estimate the efficiency or performance of CIs, techniques such as gravimetry, gasometry, adsorptiometry and electrochemistry have been used [1, 4]. The extent of protection, pitting and roughness (morphology) of the specimen surface is often checked using surface analytical techniques such as scanning electron microscopy (SEM) and atomic force microscopy (AFM), and it is usually different without and with the CI. The difference may be associated with monolayers or multilayers of adsorbed CIs on the surface. To understand the mechanism of interaction between the functionalities on the CI and the surface, various spectroscopic techniques have been used [5–7]. Such techniques include Fourier-transformed infrared spectroscopy (FTIR), atomic absorption spectroscopy (AAS), energy dispersive X-ray spectroscopy (EDS) and X-ray photoelectron spectroscopy (XPS).

Since most efficient CIs contain oxygen, nitrogen, sulphur, phosphorus, multiple bonds, aromatic and heterocyclic rings in their molecular structures [8, 9], the technique used is directed towards detecting the presence of these functionalities on the surface. The corroding environment also has to be taken into consideration. Typical corrosive environments include all forms of water, industrial, marine, neutral and urban atmospheres, air, humidity, steam, some gases like fuel gases, chlorine, hydrogen sulphide, oxides of nitrogen, ammonia and sulphur (IV) oxide, carbon dioxide, soils, alkalis and acids [8–10]. In the oilfield, acidic fluids, petroleum wastewater, brines, alkalis, microbial rich fluids and some gases constitute the corrosive environments [11]. To monitor corrosion, a selected corrosive environment may be simulated in the laboratory and placed in contact with the metal or alloy for a stipulated contact time. After this time, the metal or alloy substrate (usually in form of a coupon) is cleaned by standard procedures, and the corrosion rate can be evaluated. In this chapter, emphasis will be laid on corrosion in oilfield environments with preference to acidising (HCl) solution.

2. Corrosion in acidising solution

Acid corrosion constitutes the major source of corrosion often encountered in the oilfield. Strong acids can be pumped into wells to increase formation permeability and stimulate production. About 5–28% HCl is commonly used for limestone formations, while hydrogen fluoride (up to 3%) is used for sandstone formations [8]. Also, about 9% formic acid has been used in some deep sour wells where HCl inhibitors lose their effectiveness [1]. These acids in the live and spent forms cause serious corrosion problems.

2.1 Corrosion monitoring and measurement of corrosion rate

Steel is the main material used for construction of pipework in the oilfield. Since the main chemical constituent of steel is iron, the dissolution of iron in acid (HCl) may be illustrated using Eq. (1):

$$Fe(s) + 2HCl(aq) = FeCl_2(aq) + H_2(g) \tag{1}$$

Therefore, the corrosion of steel may be monitored using the following methods [12]:

- Determining the mass of $Fe_{(s)}$ left (or lost) after dissolution within a given time frame by weight loss (or gravimetric) technique.
- Determining the changes in resistance to charge transfer between the steel surface and the electrolyte which are embedded by electrochemical impedance spectroscopy (EIS).
- Determination of changes in activity or electrochemical potential or resistance using electrochemical techniques.
- Measuring the speed of release of hydrogen gas or the amount released within a time frame by hydrogen evolution or gasometric technique.
- Determining the amount of the Fe(s) leached into the solution at intervals by AAS or EDS and so on.

2.2 Corrosion products analyses

For any of the methods used to monitor the progress of corrosion, the steel specimen has to be immersed in the corroding medium for a given frame of time, then retrieved. Within this time, a measureable extent of corrosion would have occurred and the corrosion products will be present either in the corroding medium (from which the specimen was retrieved) and/or on the retrieved steel surface. At this point, either the solution or the surface or both may be subjected to spectroscopy or other analyses. The proceeding sections will concentrate only on the frequently used spectroscopic methods for corrosion analyses. These are: electrochemical impedance spectroscopy (EIS), ultraviolet/visible (UV-Vis) spectroscopy, FTIR spectroscopy, XRD spectroscopy, EDAX spectroscopy, AAS and XPS.

3. Spectroscopic techniques for corrosion analyses

In this section, our focus will be on the familiar experimental spectroscopic methods used to monitor corrosion and analyse corrosion products in the laboratory. Except where otherwise stated, steel is mainly considered as the substrate/specimen. Most spectroscopic techniques are associated with the interaction of the surface with electromagnetic radiation. Often, direct contact with the specimen may not be necessary. While some of the techniques can be used for *in situ* corrosion monitoring, techniques for detection and analyses of corrosion products are not associated with *in situ* analysis.

3.1 Electrochemical impedance spectroscopy (EIS)

EIS is an experimental method that is based on the measurement of impedance as a function of frequency or angular frequency of a sinusoidal perturbation of small amplitude under steady-state conditions [13]. Since frequency is used as a variable, it is termed as 'spectroscopy'. The measurement is often conducted within a large frequency range that covers several decades, from mHz to hundreds of kHz. Sometimes, the frequency range could cover μHz to MHz.

In immersion tests, the steel coupon is abraded or polished to mirror finishing, cleaned (using standard procedures such as ASTM, NACE, etc.) and soldered to the wire to obtain the working electrode. Often, only one surface (of known area) on the coupon is exposed, while the other surfaces are insulated, often using epoxy adhesives [14]. For some specialised cells, there is no need for soldering to a wire as the cell has a ready compartment (with a hole) for insertion of the coupon. In this case, the coupon 'laps' on a conductor, which connects the coupon with the other components of the cell once the electrolyte is poured into the cell.

On complete connection of the cell to the potentiostat/galvanostat, a green signal shows on the screen to signify that the cell is ready for measurement taking. With EIS, corrosion data may be obtained *in situ*. Many researchers have reported that the cell will be initially allowed to stand for some time so that corrosion can occur and open circuit potential (OCP) may stabilise [15–18]. For instance, the curves in **Figure 1a–b** show variation of OCP with standing time over a period of 30 min. It can be observed that the OCP stabilises for Specimens A and B within 120 s but is unstable for Specimen C even after 1800 s. Once the actual measurement begins, a graphical representation of the measured impedance is obtained at different frequencies. The most frequently used coordinates for presenting impedance data include:

- Plot of frequencies as a set of points in the complex-impedance plane with the real components of impedance (Z') on the x-axis and the imaginary component ($-Z''$) on the y-axis or vice versa. This plot is commonly called Nyquist plot or complex plane diagram, and an example is shown in **Figure 1c**.

- Plots of Z' and $-Z''$ as a function of frequency (usually $\log f$). This presentation is not popular for experimental cases in literature.

- Plots of magnitude of impedance /Z/ or log/Z/ and argZ against $\log f$, often referred to as Bode plot.

The measured data for $Z(w)$ are interpreted by comparison with predictions of a theoretical model or by the use of an equivalent circuit. Use of equivalent circuit is commonly reported [19, 20] and the output parameters are determined depending on the best fitting equivalent circuit. Basically, best fitting of theoretically calculated impedance plots to experimentally obtained curves affords acceptable results if there is sufficiently small deviation (often expressed as χ^2). Charge transfer resistance is obtained at the point of intercept of the curve with the impedance axis. Depending on the equivalent circuit model used, other parameters such as double-layer capacitance, solution resistance, film resistance and (constant phase element) CPE constant may be obtained from the analyses of the EIS spectrum.

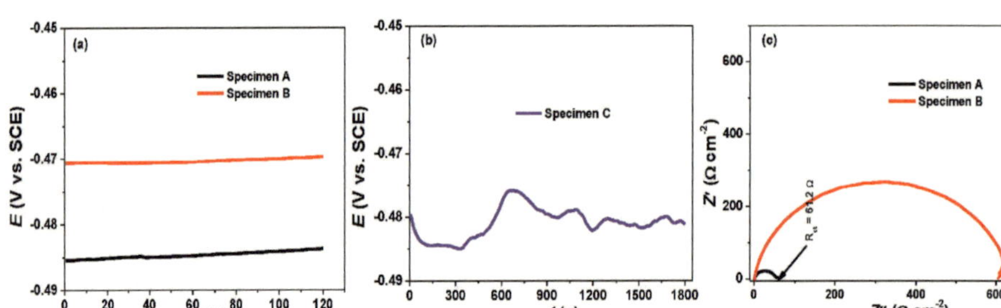

Figure 1.
(a and b) Variation of OCP with time and (c) Nyquist plot.

3.2 UV-Vis spectroscopy

Electromagnetic radiation within the ultraviolet and visible regions produces electronic excitation of molecules in a system. With the use of CIs, electrons move from one energy level to another. Depending on the chemical identity of the CI, excitations could be from the non-bonding (n) orbital (if the CI contains N, O, S and P) or from the pie (π) bonding orbitals (if it contains multiple bonds or aromatic rings) to other allowed orbitals. Such transitions are usually accompanied by the characteristic absorption in the UV-Vis region, resulting in a peak in the spectrum of the CI when scanned by the UV-Vis spectrophotometer. The wavelength(s) of maximum absorption (λ_{max}) may be recorded and used to probe corrosion. Also, formation of a complex between inhibitor species and steel species can result in another absorption and yield of similar or different λ_{max} .

In many corrosion studies, the steel specimen is immersed in the test solution for a period of time for reasonable corrosion to occur, then retrieved [8]. Aliquots of the resulting solution (containing the corrosion products) is diluted and scanned within suitable range of wavelengths [21]. Results produce absorption bands which could be used to predict the associated electronic transition. Often, the results are interpreted based on shifts in the adsorption spectra without and with the inhibitor. Typical UV-Vis spectra obtained by monitoring corrosion of X80 steel in 1 M HCl before and after immersion of steel in the inhibited solution are shown in **Figure 2a**. A shift in the spectral position in the after immersion indicates that there is interaction between the inhibitor and the steel species. Such interactions are often believed to result from the formation of a complex between active sites on the inhibitor molecule and Fe in steel [22, 23]. However, the use of UV-Vis data to predict the corrosion inhibitor-steel surface interaction is limited and considered obsolete because it does not provide explicit mechanistic information like other techniques.

3.3 FTIR spectroscopy

Vibration of functional groups on the interaction with electromagnetic radiation is the bases for application of FTIR in corrosion studies. Since each functional group has a characteristic vibrational frequency, it is easy to deploy this technique to characterise corrosion products, although not *in situ*. The surface of the steel coupon retrieved from the inhibited solution is usually cleaned, and the thin inhibitor film is gently removed. The removed film is therefore tableted with KBr and analysed by

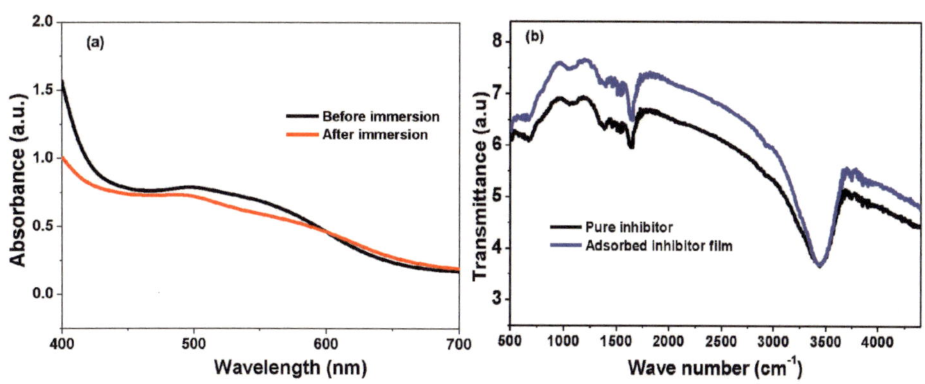

Figure 2.
(a) Typical UV-Vis spectra before and after immersion of X80 steel in 1 M HCl containing a corrosion inhibitor and (b) FTIR spectra of a pure corrosion inhibitor and adsorbed inhibitor film after immersion.

FTIR to obtain a spectrum. This is also repeated using the pure inhibitor material and another spectrum is obtained. Both spectra are often compared and shifts in position of peaks corresponding to vibrational frequencies of associated functionalities are used to predict the mechanism of interaction between the inhibitor and the steel surface [24, 25].

The experimental peaks are often matched with standard IR charts such as that of NIST or Sigma Aldrich. Example of FTIR spectra for a nanoscale bio-based corrosion inhibitor evaluated in 1 M HCl is shown in **Figure 2b**. More information on the prediction of the associated functional groups based on the obtained spectral information is provided in the following application section.

3.4 XRD spectroscopy

XRD provides information on the atomic and molecular structure of crystalline materials. It achieves this by observing the interference patterns resulting from the elastic scattering of X-rays on crystal planes. Recently being deployed for monitoring corrosion and corrosion products analyses, the technique involves scanning the surface of cleaned steel specimens retrieved from inhibited and uninhibited solutions [26]. Since Fe is the main composition of steel, the intensity of Fe obtained in the absence of the inhibitor is expected to be higher. In a recent report, Obot and co-workers observed that the intensity of Fe (100) peak obtained in the presence of inhibitors was lower than without inhibitors. They inferred that the adsorption of the CIs block off the Fe (110) reactive sites, making it less detectable during the XRD characterisation.

3.5 EDAX spectroscopy

Another X-ray technique used to detect corrosion products is the EDS. The surface of a cleaned steel specimen is scanned and the elemental composition on the surface is profiled. Without and with the corrosion inhibitor, the spectral features, percentage composition or abundance of desired elements could be compared. Comparison enables researchers to observe which element has increased in composition or has decreased. The data are therefore correlated to predict the mechanism of interaction between the surface-adsorbed inhibitor and the substrate. For instance, in **Figure 3a**, the EDS spectral features of X80 steel immersed in 1 M HCl is shown, while the one immersed in 1 M HCl containing a particular organic inhibitor (name withheld) is shown in **Figure 3b**. The differences in the spectral profiles have been used to describe the corrosion process in the following application section. Although many recent researches have deployed this method for mechanistic determinations, EDS is considered not sensitive for predicting the interactions of functional groups, except only the elemental composition.

3.6 AAS spectroscopy

When a steel coupon is immersed in a corrosive medium over a duration of time (t), the amount (or concentration) of iron leached into the medium can be determined by AAS. With this, the corrosion rate can be estimated. Suppose this is repeated in the presence of inhibitor compounds, then the corrosion inhibition efficiency can be determined in relation to the uninhibited solution [27]. The surface of the steel is prepared by following a standard method. After immersion, cleaning of the surface is not necessary as the analyses are carried out on the resulting solution containing the corrosion products. AAS presents a kind of *in situ* method for monitoring corrosion whereby aliquots of medium (during the progress of

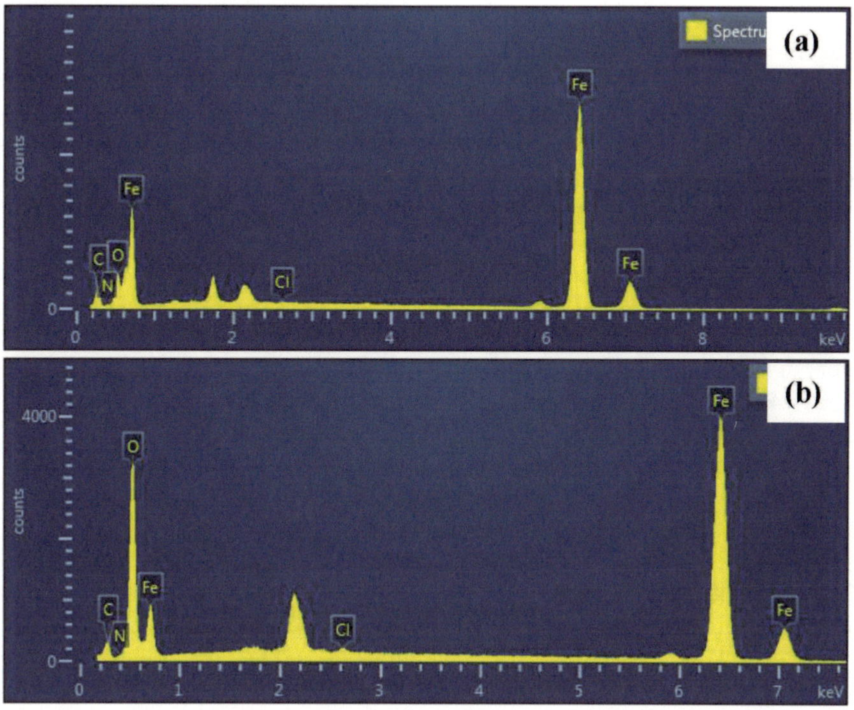

Figure 3.
EDS spectra showing the composition of Fe, C, N, O and Cl on X80 steel surface after immersion in (a) 1 M HCl and (b) 1 M HCl containing an organic corrosion inhibitor.

corrosion) may be taken at regular time interval and analysed. It could be possible to kinetically determine the order and rate constants using this technique. However, it is not often reported because it is regarded as insensitive since only the concentration of iron is used for calculation, whereas other leached metals that constitute the steel are not considered.

3.7 XPS spectroscopy

X-ray photoelectron spectroscopy or electron spectroscopy for chemical analyses (ESCA) is one of the essential tools in corrosion analyses nowadays. With XPS, the composition of a surface and the thickness of a film can be quantitatively estimated [28]. Being non-destructive on the specimen, it can also be used to probe the composition of the steel surface under the surface film. Typical escape depths of XPS experiments are estimated at ca. 3 nm at maximum, with the photoelectrons accelerated at kinetic energies of 100–1500 eV under nitrogen purging. With this, only information on thin surface region may be obtained and in the form of chemical shifts corresponding to different chemical states [29]. Results often generate two types of spectra. The first is called the survey spectrum which shows all the elements present on the specimen surface. The second is often called the deconvoluted spectrum which is used to correlate the type of bond present based on the characteristic binding energy obtained.

4. Application of spectroscopy in corrosion analyses

Spectroscopy plays an important role in corrosion analyses, especially as information on the mechanism of interaction can be derived from a single or combination of spectroscopic data. For instance, in the study of corrosion inhibition of X80 steel in 1 M HCl using L-theanine as corrosion inhibitor, the mechanism of interaction between the surface and inhibitor was described using FTIR in addition to other quantities such as binding/interaction energy [30]. FTIR results also supported quench molecular dynamics computations and provided insights into the possible adsorption sites at both molecular level and experimentally.

When a protective film is formed by a corrosion inhibitor on the surface of it its substrate, the interaction between the film and the surface is usually facilitated by some active groups in the inhibitor. Usually, the spectrum of the pure inhibitor will be different from that of its surface film. The peaks corresponding to some active groups could disappear, broaden, sharpen or shift to higher or lower wave numbers [31, 32]. The obtained spectra for pure LTN and its surface film are shown in **Figure 4** [30]. The sharp prominent peaks around 1100 cm^{-1} were assigned to C–O or C–N stretch. This peak shifted to 1055 cm^{-1}, indicating that there could have been slight modification due to its involvement in the adsorption process. Similarly, shifts and changes in intensity can be observed with peaks at 1360–1630 and 2930 cm^{-1} which were assigned to C–O stretch, C=O or N–H stretch and O–H or C–H stretch, respectively. It was inferred that such shifts signify that C=O, N–H or O–H sites may have been active in interacting with the steel surface, which could have resulted in the modification of the spectral properties. More conspicuously, the peak at 3460 cm^{-1} assigned to –N–H amine or –O–H (intermolecular bond) was found to become very broad after adsorption, a clear evidence that the sites could have been modified after taking part in the adsorption.

EDS has been widely used to profile steel surface and its corrosion products [33] in addition to FTIR. The spectrum in **Figure 3a** represents the mapped entire surface of a highly corroded surface retrieved from 1 M HCl (**Figure 5a**), whereas the spectrum in **Figure 3b** profiled the highlighted section on the surface retrieved from the inhibited solution (**Figure 5b**). To make proper meaning, the percentage compositions of the respective elements on the surface are displayed in **Table 1**.

From results presented in **Table 1**, the surface is very rich in C and Fe, which are the main elements in steel as well as O perhaps due to the oxidation of surface. This can be confirmed from the higher amounts of Fe, C and O obtained with the uninhibited system. Interaction of surface species with chloride ions in the HCl

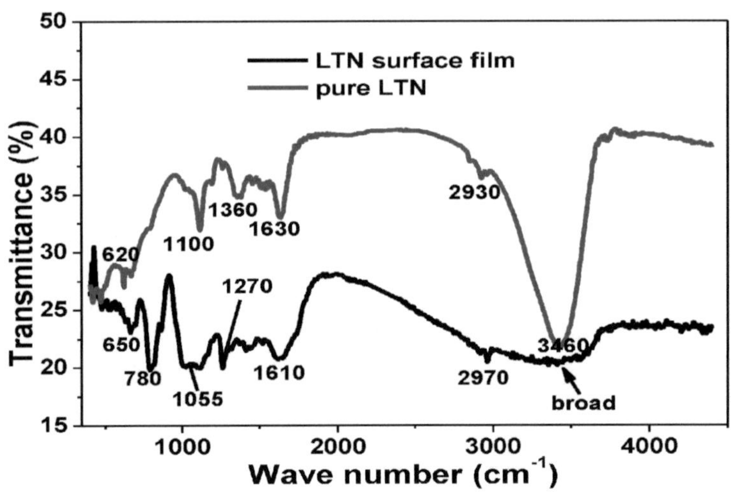

Figure 4.
FTIR spectral profiles of pure LTN, and LTN thin film formed on X80 steel surface [30].

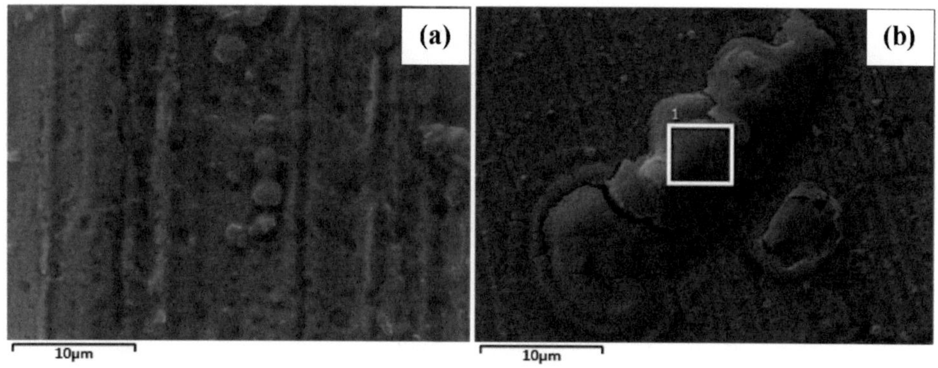

Figure 5.
Surfaces of X80 steel profiled by EDS.

Element	Uninhibited			Inhibited		
	Apparent conc.	Wt.%	Atomic %	Apparent conc.	Wt.%	Atomic %
C	5.08	12.29	26.01	1.29	2.52	8.27
N	0.03	0.12	0.18	0.07	0.26	0.23
Cl	0.46	0.39	0.44	0.23	0.04	0.13
O	12.98	0.17	12.17	33.31	13.02	32.10
Fe	150.98	87.03	61.20	89.94	84.16	59.27

*Results of ongoing research (unpublished) by Ekemini Ituen.

Table 1.
Elemental composition of the profiled X80 steel surface retrieved from (a) uninhibited and (b) inhibited 1 M HCl solution.*

medium also resulted in some amounts of Cl being found on the surface. However, on addition of the organic inhibitor which contains C=C, N and O in its molecular structure, a thin film was formed on the surface which protected it from excessive corrosive attack. This resulted in the reduction in the amounts of C, Cl and Fe 'seen' on the surface due to the 'blanketing' effect of the inhibitor. Instead, the amounts O and N on the surface increased which could have come from the adsorbed inhibitor species. In some reports, the amount of C was found to increase [34, 35], and it was attributed to involvement of C=C sites in the adsorption process.

During the evaluation of a benzimidazole derivative (BPMB) as alternative carbon steel corrosion inhibitor in CO_2-saturated brine solution, the steel surfaces in the uninhibited and inhibited media were analysed by XRD [26]. Obot and co-workers report that the most prominent diffraction peaks were observed at $2\theta = 44.9°$ and $2\theta = 65°$ assigned to Fe (110) and Fe (200) crystallographic phases. Fe (110) plane, being the predominantly oriented and stable plane in steel, showed the highest frequency in the absence of BPMB as shown in **Figure 6a**. On introduction of BPMB to the corroding medium, the intensity of the Fe (110) peak diminished (**Figure 6b**) due to adsorption of BPMB. It was implied that the adsorption of corrosion inhibitors blocked the Fe (110) reactive sites, making it less detectable during XRD characterisation [26].

In **Figure 7a**, a survey spectrum of X80 steel retrieved from 1 M HCl solution and inhibited by the plant extract-mediated silver nanoparticles as analysed by XPS is shown. Analyses of the spectrum with prejudice to Fe shows the presence of C (1s), N (1s), Ag ($^3d_5$), O (1s) at atomic percentages of 68.45, 0.64, 0.14 and 30.76%,

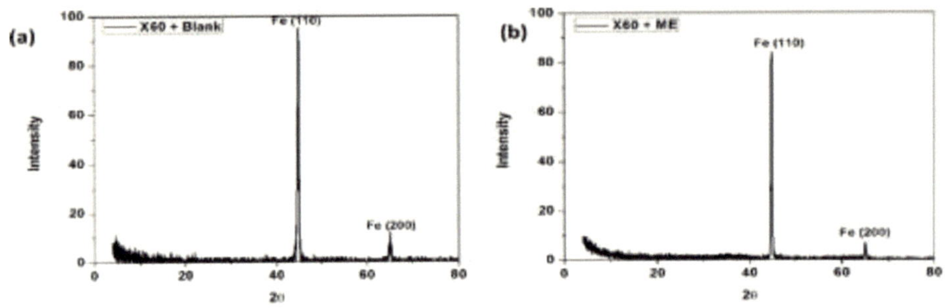

Figure 6.
XRD diffraction pattern of X60 steel retrieved from (a) uninhibited and (b) inhibited acid solution [26].

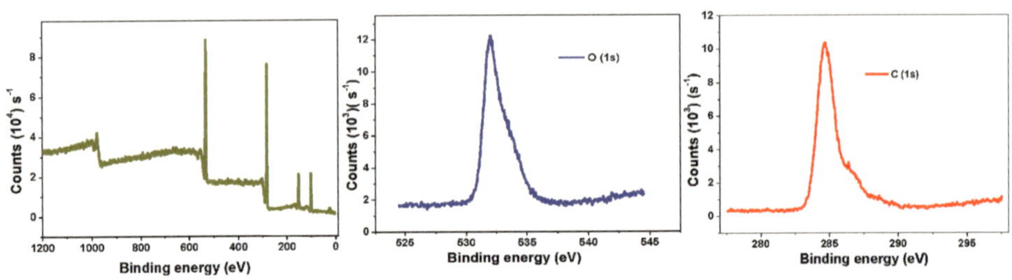

Figure 7.
(a) Survey spectra for the adsorption of onion peel extracts and silver nanoparticles on X80 steel surface in 1 M HCl, and representative spectra for (b) O (1s) and (c) C (1s).

respectively. However, the survey spectrum of the steel surface in the absence of the inhibitor (not shown) yielded no Ag and no N atoms and also lower percentages of C and O on the surface. These demonstrate that the silver nanoparticles were adsorbed on the steel surface, hence the appearance of Ag. The Ag was coordinated to phyto-compounds from the plant extract possibly by means of C=C, C–O, C–N, O–H and C=O sites which enhanced their adsorptive binding on the steel surface. The deconvoluted spectra are illustrated for the C and O associated with the surface as shown in **Figure 7b** and **c**.

Recently, plant-mediated nickel nanoparticles (Ni-EtNPs) have been demonstrated as highly efficient corrosion inhibitor for X80 steel in 1 M HCl [4]. Electrochemical impedance spectroscopy (EIS) was used to probe the corrosion progress and kinetics. It was observed that it took between 300 and 1200 s for OCP to stabilise (**Figure 8a**) depending on the test solution studied. Nyquist plots (**Figure 8b**) yielded depressed semicircles whose diameters increased as concentration of the Ni-EtNPs increased. All the curves showed one relaxation time constant, demonstrating that corrosion mechanism was the same charge transfer-controlled corrosion process in all the test solutions. As Ni-EtNPs increased, the diameter of the depressed semicircle increased corresponding to an increase in charge transfer resistance. The increase in charge transfer resistance was attributed to be caused by adsorption of a thin film of Ni-EtNPs on the steel surface, and hence corrosion inhibition. The possible scenario at the electrode-film-solution interfaces was modelled by using the equivalent circuit, which is composed of the solution resistance (R_s), film resistance (R_f), charge transfer resistance (R_{ct}) and two constant phase elements (CPEs). On analyses using Gamry E-Chem software, the associated parameters (**Table 2**) were obtained.

The obtained values of R_{ct} increased as concentration of Ni-EtNPs increased signifying that the coverage, number of adsorbed species and corrosion inhibition

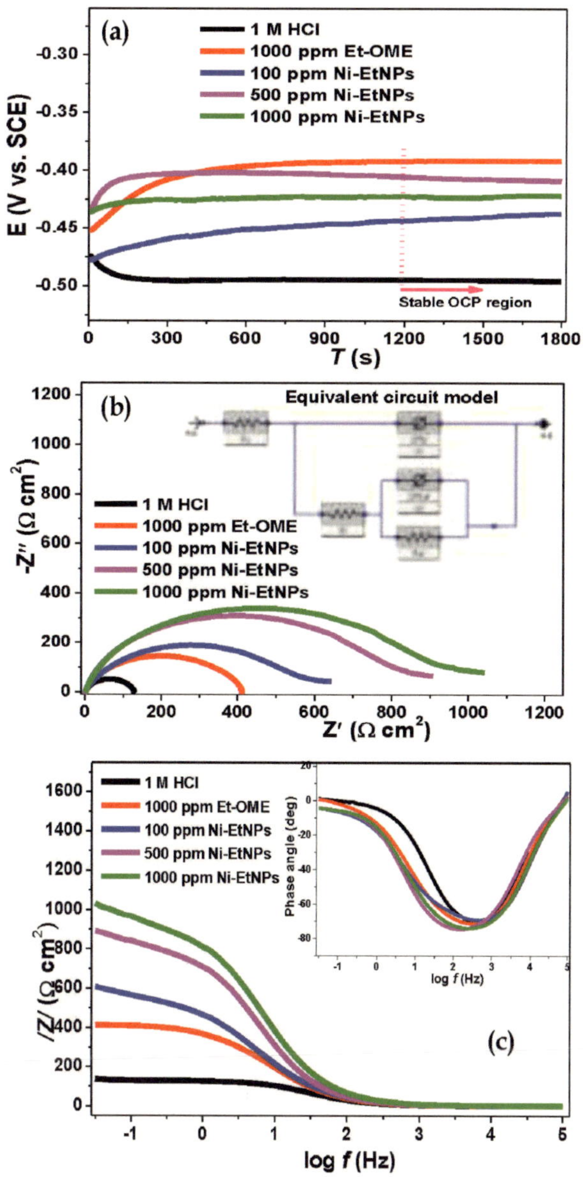

Figure 8.
(a) OCP – Time plot, (b) Nyquist plot with equivalent electrical circuit embedded and (c) bode modulus plot with phase angle plot embedded for the corrosion of X80 steel in 1 M HCl inhibited by various concentrations of Ni-EtNPs at 25°C [4].

increased as concentration increased. The adsorbed Ni-EtNPs species were considered to have formed a surface film with resistance denoted by R_f, which also increased with increase in concentration. By implication, there was increase in the local dielectric or insulation within the region of the surface as a result of adsorption of inhibitor film. As more inhibitor molecules were adsorbed, water molecules were being displaced from the surface leading to decrease in active corrosion sites and reduced corrosion rate. Bode and phase angle plots extracted from obtained impedance data are shown in **Figure 8c**. Values of resistances obtained were consistent with the serial arrangement $R = R_f + R_{ct}$ and were compared to using a crude extract as corrosion inhibitor.

Test solution	R_s (Ωcm^2)	CPE_f		CPE_{dl}				
		n_1	$R_f(\Omega cm^2)$	n_2	$R_{ct}(\Omega cm^2)$	$R(\Omega cm^2)$	χ^2 **(x10^{-5})**	$I_{EIS}(\%)$
1 M HCl	0.84 ± 0.12	0.84	1.6 ± 0.2	0.70	130.7 ± 1.4	130.2 ± 1.6	0.7	—
1000 ppm Et-OME	0.86 ± 0.10	0.86	11.3 ± 0.2	0.71	389.5 ± 1.6	409.8 ± 2.3	0.8	67.2
100 ppm Ni-EtNPs	0.98 ± 0.11	0.86	17.4 ± 0.2	0.71	686.0 ± 1.4	703.4 ± 2.1	1.5	80.9
500 ppm Ni-EtNPs	0.91 ± 0.15	0.87	24.2 ± 0.3	0.73	954.5 ± 3.2	988.7 ± 2.6	2.9	86.3
1000 ppm Ni-EtNPs	1.08 ± 0.12	0.88	35.1 ± 0.3	0.75	1106.5 ± 3.6	1141.6 ± 4.8	6.2	88.1

Table 2.
EIS parameters for corrosion of X80 steel in 1 M HCl and inhibited by Et-OME and Ni-EtNPs [4].

5. Conclusion

Spectroscopic techniques play key roles in determining monitoring corrosion, analysing corrosion products and predicting mechanism of inhibitor-surface interactions. EIS provides *in situ* information on corrosion rate, corrosion mechanism, charge transfer resistance and double-layer kinetics. UV-Vis provides information on possible formation of a complex between the inhibitor and metal ions from the steel surface signalled by electronic transitions. FTIR and EDS afford information on possible active functionalities on the inhibitor that binds at the surface's active sites, backed up by XPS results. Thus, a combination of the various spectroscopic techniques discussed could be superb for characterising the inhibition of corrosion of steel by a given inhibitor in the acidic oilfield environment.

Author details

Ekemini Ituen[1,2,3], Lin Yuanhua[1,2*], Ambrish Singh[1,2*] and Onyewuchi Akaranta[4]

1 State Key Laboratory of Oil and Gas Reservoir Geology and Exploitation, Southwest Petroleum University, Chengdu, Sichuan, China

2 School of Materials Science and Engineering, Southwest Petroleum University, Chengdu, Sichuan, China

3 Emerging Materials and Energy Research Group, Department of Chemistry, University of Uyo, Uyo, Nigeria

4 African Center of Excellence for Oilfield Chemicals Research, University of Port Harcourt, Port Harcourt, Nigeria

*Address all correspondence to: yhlin28@163.com and vishisigh4uall@gmail.com

IntechOpen

© 2020 The Author(s). Licensee IntechOpen. This chapter is distributed under the terms of the Creative Commons Attribution License (http://creativecommons.org/licenses/by/3.0), which permits unrestricted use, distribution, and reproduction in any medium, provided the original work is properly cited.

References

[1] Finšgar M, Jackson J. Application of corrosion inhibitors for steels in acidic media for the oil and gas industry: A review. Corrosion Science. 2014;**86**:17-41

[2] Kokalj A, Peljhan S, Finsgar M, Milosev I. What determines the inhibition effectiveness of ATA, BTAH, and BTAOH corrosion inhibitors on copper? Journal of the American Chemical Society. 2010;**132**(46):16657-16668

[3] Sastri VS. Types of corrosion inhibitor for managing corrosion in underground pipelines. In: Underground Pipeline Corrosion. Amsterdam: Woodhead Publishing; 2014. pp. 166-211

[4] Ituen E, Singh A, Yuanhua L. Inhibitive effect of onion mesocarp extract-nickel nanoparticles composite on simultaneous hydrogen production and pipework corrosion in 1 M HCl. International Journal of Hydrogen Energy. 2020;**45**:10814-10825

[5] Ituen E, Mkpenie V, Yuanhua L, Singh A. Inhibition of erosion corrosion of pipework steel in descaling solution using 5-hydroxytryptamine-based additives: Empirical and computational studies. Journal of Molecular Structure. 2020;**1204**:127562

[6] Ituen EB, Akaranta O, Umoren SA. N-acetyl cysteine based corrosion inhibitor formulations for steel protection in 15% HCl solution. Journal of Molecular Liquids. 2017;**246**:112-118

[7] Fabis PM, Brown CW, Rockett T, Heidersbach RH. Infrared and Raman spectroscopic analysis of corrosion products on metals exposed to hydrogen sulfide at elevated temperatures. Journal of Materials for Energy Systems. 1981;**3**(2):66-75

[8] Ahmad Z. Principles of Corrosion Engineering and Corrosion Control. Amsterdam: Elsevier; 2006

[9] Kermani MB, Morshed A. Carbon dioxide corrosion in oil and gas production—A compendium. Corrosion. 2003;**59**(8):659-683

[10] Ren C, Liu D, Bai Z, Li T. Corrosion behavior of oil tube steel in simulant solution with hydrogen sulfide and carbon dioxide. Materials Chemistry and Physics. 2005;**93**(2-3):305-309

[11] Iverson WP. Microbial corrosion of metals. In: Advances in Applied Microbiology. Vol. 32. Cambridge, MA: Academic Press; 1987. pp. 1-36

[12] Choudhry KI, Carvajal-Ortiz RA, Kallikragas DT, Svishchev IM. Hydrogen evolution rate during the corrosion of stainless steel in supercritical water. Corrosion Science. 2014;**83**:226-233

[13] Bard AJ, Inzelt G, Scholz F, editors. Electrochemical Dictionary. New York, NY: Springer Science & Business Media; 2008

[14] Ituen E, Mkpenie V, Ekemini E. Corrosion inhibition of X80 steel in simulated acid wash solution using glutathione and its blends: Experimental and theoretical studies. Colloids and Surfaces A: Physicochemical and Engineering Aspects. 2019;**578**:123597

[15] Ituen E, Ekemini E, Yuanhua L, Singh A. Green synthesis of Citrus reticulata peels extract silver nanoparticles and characterization of structural, biocide and anticorrosion properties. Journal of Molecular Structure. 2020;**1207**:127819

[16] Xu X, Singh A, Sun Z, Ansari KR, Lin Y. Theoretical, thermodynamic and electrochemical analysis of biotin drug

as an impending corrosion inhibitor for mild steel in 15% hydrochloric acid. RSC Advances. 2017;**4**:170933. DOI: 10.1039/C9NJ01691K

[17] Obot IB, Ankah NK, Sorour AA, Gasem ZM, Haruna K. 8-Hydroxyquinoline as an alternative green and sustainable acidizing oilfield corrosion inhibitor. Sustainable Materials and Technologies 2017;14:1-0.

[18] Ituen EB, James AO, Akaranta O. Fluvoxamine-based corrosion inhibitors for J55 steel in aggressive oil and gas well treatment fluids. Egyptian Journal of Petroleum. 2017;**26**(3):745-756

[19] Yuan X-Z, Song C, Wang H, Zhang J. Electrochemical Impedance Spectroscopy in PEM Fuel Cells. Springer London; 2010. DOI: 10.1007/978-1-84882-846-9

[20] Jia WA, Mengyang JI, Zhaohui YA, Bing HA. On completeness of EIS equivalent circuit analysis for electrochemical corrosion process. Journal of the Chinese Society of Corrosion and Protection. 2018;**37**(6):479-486

[21] Li Y, He JB, Zhang M, He XL. Corrosion inhibition effect of sodium phytate on brass in NaOH media. Potential-resolved formation of soluble corrosion products. Corrosion Science. 2013;**74**:116-122

[22] Vinutha MR, Venkatesha TV. Review on mechanistic action of inhibitors on steel corrosion in acidic media. Portugaliae Electrochimica Acta. 2016;**34**(3):157-184

[23] Shainy KM, Ammal PR, Unni KN, Benjamin S, Joseph A. Surface interaction and corrosion inhibition of mild steel in hydrochloric acid using pyoverdine, an eco-friendly bio-molecule. Journal of Bio- and Tribo-Corrosion. 2016;**2**(3):20

[24] Alaneme KK, Olusegun SJ, Adelowo OT. Corrosion inhibition and adsorption mechanism studies of Hunteria umbellata seed husk extracts on mild steel immersed in acidic solutions. Alexandria Engineering Journal. 2016;**55**(1):673-681

[25] Chitra S, Anand B. Surface morphological and FTIR spectroscopic information on the corrosion inhibition of drugs on mild steel in chloride environment. Journal of Chemical and Pharmaceutical Sciences. 2017;**10**:453-456

[26] Obot IB, Onyeachu IB, Umoren SA. Alternative corrosion inhibitor formulation for carbon steel in CO2-saturated brine solution under high turbulent flow condition for use in oil and gas transportation pipelines. Corrosion Science. 2019;**159**:108140

[27] Ituen E, Akaranta O, James A. Green anticorrosive oilfield chemicals from 5-hydroxytryptophan and synergistic additives for X80 steel surface protection in acidic well treatment fluids. Journal of Molecular Liquids. 2016;**224**:408-419

[28] Asami K, Hashimoto K. X-ray photoelectron spectroscopy for corrosion studies. Langmuir. 1987;**3**(6):897-904

[29] Walczak MS, Morales-Gil P, Belashehr T, Kousar K, Lozada PA, Lindsay R. Determining the chemical composition of corrosion inhibitor/metal interfaces with XPS: Minimizing post immersion oxidation. Journal of Visualized Experiments. 2017;**121**:e55163

[30] Ituen E, Mkpenie V, Dan E. Surface protection of steel in oil well acidizing fluids using L-theanine-based corrosion inhibitor formulations: Experimental and theoretical evaluation. Surface and Interface Analysis. 2019;**16**:29-42

[31] Salarvand Z, Amirnasr M, Talebian M, Raeissi K, Meghdadi S. Enhanced corrosion resistance of mild steel in 1 M HCl solution by trace amount of 2-phenylbenzothiazole derivatives: Experimental, quantum chemical calculations and molecular dynamics (MD) simulation studies. Corrosion Science. 2017;**114**:133-145

[32] Ameh PO. A comparative study of the inhibitory effect of gum exudates from *Khaya senegalensis* and *Albizia ferruginea* on the corrosion of mild steel in hydrochloric acid medium. International Journal of Metals [Internet]. Hindawi Limited. 2015;**2015**:1-13. DOI: 10.1155/2015/824873

[33] Ituen E, Akaranta O, James A. Electrochemical and anticorrosion properties of 5-hydroxytryptophan on mild steel in a simulated well-acidizing fluid. Journal of Taibah University for Science. 2017;**11**(5):788-800

[34] Ituen E, Ekemini E, Yuanhua L, Li R, Singh A. Mitigation of microbial biodeterioration and acid corrosion of pipework steel using Citrus reticulata peels extract mediated copper nanoparticles composite. International Biodeterioration and Biodegradation. 2020;**149**:104935

[35] Ituen E, Singh A, Li R, Yuanhua L, Guo C. Nanostructure, surface and anticorrosion properties of phyto-fabricated copper nanocomposite in simulated oilfield descaling fluid. Surface and Interface Analysis. 2020;**3**:100514

Chapter 8

Kinetics and Structure Aspects of the Dissolution of Stainless Steels in Phosphoric Acid

Michael Schorr, Benjamín Valdez, Margarita Stoytcheva and Roumen Zlatev

Abstract

The dissolution of metals in acids is a heterogeneous electrochemical reaction, taking place at the interface between a solid and a liquid, described in terms of energy and configuration of the species involved. The energy of activation for two stainless steels (SSs) in phosphoric acid (PA) was determined from their corrosion rates in the temperature range of 20–120°C, applying the Arrhenius equation. Energy of activation values for the overall process of anodic metal dissolution and cathodic hydrogen release is in the 10–20 Kcal mol^{-1} range. Metal corrosion is a structure-sensitive process; its mechanism and rate are related to the density of surface defects, based on the Boltzmann expression. A model for the activated complex is proposed based on considerations of energy and configuration. This work is an integration of a theoretical analysis and an experimental investigation of metal corrosion in acids.

Keywords: phosphoric acid, stainless steel, dissolution, kinetics, surface structure

1. Introduction

Acids are broadly applied in many industries: chemical, fertilizer, steel, mineral, water, oil, food, etc. [1]. The acidic properties of their ionized, aqueous solutions are due to the presence of hydrogen ion H^+ and its complexed form hydronium ion H_3O^+.

Phosphoric acid (PA), H_3PO_4, is an industrial acid used as an intermediate in the fertilizer industry, for metal surface treatment, and as an additive in the food and beverage industry [2–4]. PA, a polyprotic acid, was selected for this study since it is a nonreductant, nonoxidant acid, to avoid the complication of oxidation and formation of oxides on the metallic surface.

Corrosion rates (r) of austenitic stainless steels (SS) in concentrated PA, in the presence and in the absence of halides, have been determined by weight loss and the corrosion behavior by electrochemical measurements, in the 50–120°C temperature range.

The dissolution of metals in solutions of acids is a heterogeneous electrochemical reaction. Initially, two phases coexist—a solid and a liquid—but as the reaction starts, a third phase appears, hydrogen gas, evolving from the surface of the metal. The kinetics of the dissolution process depends on the collision between the solvated protons and the "active sites" on the metallic crystal surface, promoting anodic dissolution. The dissolution rate increases with temperature in accordance with the empirical Arrhenius equation:

$$r = A\ \exp.(-E_a/RT) \tag{1}$$

where r is the metal dissolution reaction rate, A the frequency factor, T the absolute temperature, R the gas constant in calories, and E_a the activation energy. According to the rate process theory [5–7], any process may be described in terms of energy and configuration of the species involved. In the transition from the unstable metallic crystal-acid system to the stable system of solvated metallic cations and anions, the reaction passes through the activated complex which is in dynamic equilibrium with the reactants:

$$\underset{\text{metal crystal}}{M} + \underset{\text{solvated acid}}{HA} \leftrightarrow \underset{\text{activated complex}}{[X]^*} \rightarrow \underset{\text{solvated salt}}{MA} + \underset{\text{hydrogen gas}}{\tfrac{1}{2}H_2\uparrow} \tag{2}$$

For iron in orthophosphoric acid, the overall reaction is:

$$Fe + 2H_3PO_4 \rightarrow Fe(H_2PO_4)_2 + H_2\uparrow \tag{3}$$

The reaction rate, in terms of enthalpy of activation (ΔH^*) and entropy of activation (ΔS^*), is expressed as:

$$r = (RT/Nh)\ \exp.(-\Delta H*/RT)\ \exp.(\Delta S*/R) \tag{4}$$

where N is the Avogadro number and h is the Planck constant.

2. Experimental

Laboratory immersion corrosion tests were performed in accordance with the practices recommended in ASTM Volume 03.02 [8]. ASTM G 31 standard [9] was applied for mass loss tests, and the electrochemical parameters were measured and reported following ASTM G 3 [10] and G 5 [11] standards. Reagent grade PA and halide acids were utilized, for preparing PA solutions. Two SS, currently used in PA applications, were tested; their chemical composition is shown in **Table 1**.

Corrosion rates of austenitic SS in concentrated PA, in the presence and in the absence of halides, have been determined by weight loss measurements in the 50–120°C temperature range. The data obtained were plotted in accordance with the Arrhenius equation:

$$\log\ r = \log A – E_a/(2.3RT) \tag{5}$$

Plots of log r vs. 1/T yield satisfactory straight lines and the values of E_a lie in the range of 10–22 Kcal mol^{-1}, the particular value depending on acid concentration and content of halide.

UNS*	Chemical composition, %			
	Cr	Ni	Mo	C_{max}
S 30400	18–20	8–12	—	0.08
S 31600	16–18	10–14	2–3	0.10

Table 1.
Austenitic stainless steels tested.

UNS	Acid concentration, %	Temperature, °C	Potential, V			Current density, mA cm^{-2}	
			E_{ocp}	E_{pp}	E_p	i_{crit}	i_p
S 30400	70	20	0.14	—	—	—	—
S 31600	70	20	0.15	—	—	—	—
S 30400	70	80	0.17	—	—	—	—
S 30400	85	120	−0.27	−0.04	0.22	12	0.14
S 31600	85	120	−0.19	−0.14	−0.05	0.74	0.20
S 30400	95	120	−0.27	−0.12	−0.05	2.0	0.002

Table 2.
Electrochemical parameters of stainless steel—H_3PO_4.

The kinetics of dissolution was studied also by electrochemical methods. Measurements were made of electrode potential at zero current as a function of immersion time, and the potential-current relationship was determined by potentiostatic and potentiodynamic techniques.

1. The concentrations of the PA tested are 70, 100, and 85% (see **Figure 2** and **Table 2**), and pure PA prepared from PA grade analytical reagent distilled water was used.

2. The halides tested were F, Cl, and Br.

3. Their concentration in PA was 1000 ppm.

3. Results and discussion

3.1 Kinetic and thermodynamic studies

The effect of temperature on the corrosion rate, in accordance with the Arrhenius equation, is shown in **Figure 1**.

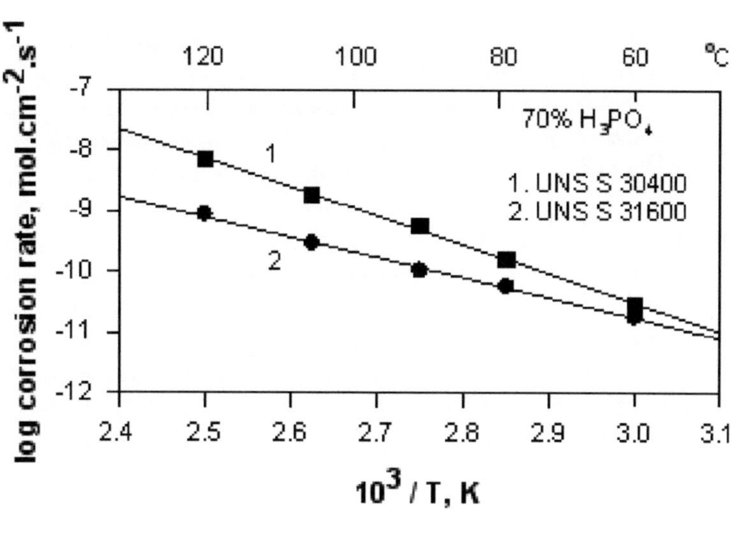

Figure 1.
Variation of the corrosion rate of UNS S 30400 and S 31600 with temperature in PA.

The enthalpy of activation ΔH* and the entropy of activation ΔS* were calculated by comparison of equations (1) and (4). The free energy of activation ΔG* was calculated from the expression:

$$\Delta G* = \Delta H* - T\Delta S* \qquad (6)$$

Selected kinetic and thermodynamic values are illustrated in **Table 3**. This table also shows the influence of the halide ions F-, Cl-, and Br- on the corrosion rate, which was added to PA as halide acids: HF, HCl, and HBr. The halide concentration in PA is 1000 ppm.

The electrochemical behavior of several steels in PA solutions is graphically illustrated by current density-potential plots (**Figure 2**). The following electrochemical parameters are located: open-circuit potential (E_{opc}), critical current density (i_{crit}), primary passive potential (E_{pp}), passivation current density (i_p), passivation potential (E_p), and transpassive potential (E_t) [10, 11].

The corrosivity of pure phosphoric acid solutions increases with the increase in concentration to about 60% PA, and then it is reduced with further increase in concentration. This behavior parallels the H+ concentration, which increments with the increase in acid concentration to about 50% PA and then decreases as the

UNS		S 30400				S 31600		
Halide	—	F^-	Cl^-	Br^-	—	F^-	Cl^-	Br^-
Corrosion rate (mdd)*	110	225	2600	110	25	20	70	10
E_a, Kcal mol^{-1}	22.3	16.8	12.6	10.9	15.4	11.6	15.5	12.2
Activated atoms, atoms cm^{-2}	18.0	4.2×10^4	1.8×10^7	2.1×10^8	3.2×10^5	7.4×10^7	2.8×10^5	1.6×10^5
ΔS*, cal $deg.^{-1}$ mol^{-1}	42.0	55.6	63.0	75.0	64.5	75.8	62.1	64.0
ΔG, Kcal mol^{-1}	36.4	35.7	34.1		37.5	37.7	36.8	34.1

Table 3.
Kinetic and thermodynamic values for corrosion of stainless steel in 70% H_3PO_4 containing 1000 ppm halide at 80°C.

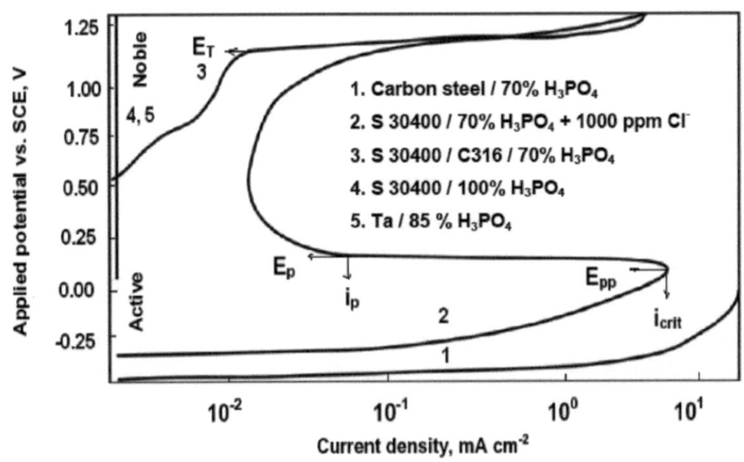

Figure 2.
Anodic potentiodynamic polarization plots in H_3PO_4 at 20°C, showing typical corrosion behavior.

equilibrium in concentrated acid shifts to the left with further increase in acid concentration [12]:

$$H_3PO_4 \leftrightarrow H^+ + H_2PO_4^- \quad (7)$$

3.2 The energy of activation

The values of activation energy obtained for different SS-PA systems lie between 10 and 22 Kcal mol^{-1}; the values of activation entropy are in the range of −42 to −75 cal deg^{-1} mol^{-1} and those of free energy of activation in the range of 34–38 Kcal mol^{-1}.

The energy of activation for the overall process of cathodic release of hydrogen from acids on different metals is in the 10–20 kcal mol^{-1} range [12, 13]. This indicates that in the dissolution of steel in phosphoric acid, and probably in metals dissolving in strong mineral, non-oxidizing acids, the hydrogen evolution reaction constitutes the rate determining step of the overall dissolution reaction:

$$M + 2H^+ \rightarrow M^{2+} + H_2\uparrow \quad (8)$$

3.3 The influence of halides on corrosion

The corrosion activity of S 30400 and S 31600 in 70% PA starts at about 60°C and 80°C, respectively [14]. Up to these temperatures, they exhibit a passive state, but with increase of temperature, breakdown of passivity occurs. The potential becomes active and an intense corrosion process is established. The values of electrochemical parameters indicating corrosion in halide-containing PAs change, in accordance with their respective corrosion rates (**Tables 2** and **3**).

Halogen acids HX were added to PAs to augment their corrosiveness; in particular, addition of HCl results in markedly increased corrosion rates. The corrosivity of these halogen acids and their ions, in strong mineral oxygen acids such as PA, is related to the halogen physicochemical properties, their electronegativity, ion size, and ionic character of the HX molecule, which indicate their chemical reactivity. Cl^- ion is adsorbed on metal surfaces and replaces adsorbed oxygen or water molecules. This shifts the metal potential to more active (electronegative) values and causes breakdown of the passive state, mainly in concentration higher than 70% at 120°C (**Table 2**). During attack on SS, chlorides of Fe, Ni, and Cr are formed; they are highly soluble in PA because of its complexing cations of the transition group elements.

The corrosivity of halide ions diminishes in the order $F^- > Cl^- > Br^-$ in 30% and 50% PAs in accordance with their chemical activity. In 70% and 85% PAs, however, the corrosivity of halides is in the order $Cl^- > F^- > Br^-$. This result reflects the formation of stable monofluorophosphoric acid, which complexes F^- and decreases its corrosivity. The following equilibrium is established:

$$H_3PO_4 + HF \leftrightarrow H_2PO_4F + H_2O \quad (9)$$

3.4 Electrochemical corrosion behavior

Typical types of corrosion behavior shown by anodic polarization plots (1)–(4), locating their electrochemical parameters, are displayed in **Figure 2**.

Active corrosion behavior of carbon steel is shown in plot 1; plot 2 traverses three regions of corrosion behavior: active, passive, and transpassive, the last indicated by the appearance of a yellow color in the colorless PA solution, due to the presence of CrO_4^{2-}, generated in the oxidative dissolution of SS and corroborated by chemical analysis. Plot 3 starts with passivity; transpassivity appears at E_t about 1.00 V, but a total passive state is maintained in 100% PA (plot 4). The electrochemical behavior of SS in PA is related to acid concentration and temperature. The presence of halide ions in PA accelerates the corrosion, the increase in rate being dependent upon the halide chemical nature and content (**Tables 2** and **3**).

Ta is used for equipment handling hot, concentrated PA due to its high corrosion resistance.

3.5 The activated complex structure

Based on considerations of energy and configuration using the values of enthalpy, entropy, and free energy of activation from equation (6), the structure of the activated complex may be described as an association between the atoms undergoing corrosion on the metal surface and the hydronium ions. This associated configuration conforms to the negative and relatively high values of the entropy of activation which is typical of chemical reactions involving dimerization or association of reactant molecules. This conclusion concurs with the fact that protons and corrosion-promoting ions such as chloride are adsorbed on the metal surface and accelerate the corrosion process. Furthermore, these values are characteristic of slow reactions such as the corrosion of stainless steels in non-oxidizing acids.

Taking into account these considerations, a model for the activated complex of the overall metal dissolution reaction is proposed:

$$MM + zH_3O^+ \Leftrightarrow [MM-(OH_3)_z^{z+}]^* \rightarrow M + M(OH_2)_z^{z+} + zH\bullet$$

metal surface — hidronium ion — activated complex — nascent metal surface — hydrated metal ion — adsorbed hydrogen atom

(10)

$$2H_{ad} \rightarrow H_2\uparrow \quad (11)$$

The positive values calculated for the free energy of activation (6) are a measure of the energy required to form the activated complex and its instability. On the other hand, in the corrosion process:

$$\Delta G = \Delta H - T\Delta S \quad (12)$$

where ΔH is the heat of reaction of an exothermic reaction and therefore negative, and ΔS is the entropy of the irreversible corrosion process, which results in an increase in the degree of disorder and therefore is positive. Consequently, the free energy (ΔG) has a negative value expressing the spontaneity of the corrosion process of metal in acids and its acceleration with the increase in temperature.

3.6 Activation and active sites

Experimental values of activation energy for metal dissolution in acid solutions are frequently reported in the literature. For instance, for Ni dissolution [15], the values lie between 10.5 and 21.1 kcal mol^{-1}, and values for other metals are similar [16–20]. It is interesting to speculate why no greater values have been reported. The answer can be found from consideration of the Boltzmann expression [18]:

$$n_a/n = \exp.[-E_a/(RT)] \tag{13}$$

where n is the total number of metal atoms per surface area unit and n_a the number of "activated" metal atoms having energies greater than E_a and therefore able to undergo dissolution. In a metallic crystal, the number of atoms per cm^2 of surface area may be taken as ca. 10^{15} [18], hence:

$$\log n_a = 15\text{–}E_a/(2.3\ RT) \tag{14}$$

For instance, at 100°C,

$$\log n_a = 15\text{–}E_a/1720 \tag{15}$$

and if E_a is 25.8 kcal mol^{-1}, n_a = 1 at cm^{-2}. Since the density of active sites is 10^8–10^{12} cm^{-2} for different cases [18], E_a values of about 25 kcal mol^{-1} or greater are unreasonable.

Metallic corrosion is considered as a structure-sensitive process, and its mechanism and rate have been related to the density of surface imperfections or defects [18, 21–26]. Continuing with calculations based on the Boltzmann expression, the density of active sites n_a in a metal undergoing dissolution with an activation energy of 10.0 kcal mol^{-1} is 1.3×10^9 at cm^{-2}, and it is 1.7×10^6 at cm^{-2} when the activation energy is 15.0 kcal mol^{-1}. The similarity between these n_a values and the density of defects in metals—for instance, 10^6 to 10^8 dislocations cm^{-2} in a well-annealed metal [26–28]—strongly suggests a quantitative relationship between metal dissolution and density of surface defects.

4. Conclusion

Corrosion rates and electrochemical behavior of austenitic stainless steels UNS S 30400 and S 31600 in PA solutions, with and without halide ions, have been determined in the 20–120°C temperature range. The energy of activation E_a for the dissolution of these SS in PA solutions was obtained in accordance with the Arrhenius equation. The values of E_a for different SS-PA systems lie between 10 and 20 kcal mol^{-1}. Calculations based on the Boltzmann expression, taking into account the activation energy, the total number of metal atoms, and the number of "activated" metal atoms, both per surface area unit, strongly suggest a quantitative relationship between metal dissolution and density of defects in metal's surface. Based on considerations of energy and configuration, using the values of energy, entropy, and free energy of activation, a model for the activated complex of the overall metal dissolution reaction in acids is proposed.

Author details

Michael Schorr*, Benjamín Valdez, Margarita Stoytcheva and Roumen Zlatev
Universidad Autónoma de Baja California, Instituto de Ingeniería, Mexicali, México

*Address all correspondence to: mschorr2000@yahoo.com

IntechOpen

© 2020 The Author(s). Licensee IntechOpen. This chapter is distributed under the terms of the Creative Commons Attribution License (http://creativecommons.org/licenses/by/3.0), which permits unrestricted use, distribution, and reproduction in any medium, provided the original work is properly cited. CC BY

References

[1] Dillon CP. Corrosion Control in the Chemical Process Industries. 2nd ed. MTI, NACE International; 1994

[2] Lehmann U. Environmentally sustainable recycling of acids for pickling of stainless steel. In: ThysenKrupp Techforum; 2006. p. 26

[3] Schorr M. Stainless steel for corrosion control in the phosphoric acid industry. Stainless Steel World. 1998;**10**:25

[4] Ross RW. In: Cramer SD, Covino Jr. BS, editors. ASM Handbook. Vol. 13C. Materials Park, Ohio: ASM; 2006. p. 736

[5] Glasstone S, Laidler RJ, Eyring H. The Theory of Rate Process. London: McGraw Hill Book; 1941

[6] Vetter KJ. Electrochemical Kinetics. NY: Academic Press; 1967. p. 134

[7] Stansbury EE, Buchanan RA. Fundamentals of Electrochemical Corrosion. Materials Park, Ohio: ASM International; 2000. p. 91

[8] ASTM. 2006 Annual Book of ASTM International Standards, ASTM, Vol. 03.02. Wear and Erosion; Metal Corrosion. West Conshohocken, PA: ASTM; 2006

[9] ASTM. ASTM G 31, Laboratory Immersion Corrosion Testing of Metals, ASTM Vol. 03.02. West Conshohocken, PA: ASTM; 2006

[10] ASTM. ASTM G 3, Conventions Applicable to Electrochemical Measurements in Corrosion Testing, ASTM Vol. 03.02. West Conshohocken, PA: ASTM; 2006

[11] ASTM. ASTM G 5, Method for Making Potentiostatic and Potentiodynamic Anodic Polarization Measurements, ASTM Vol. 03.02. West Conshohocken, PA: ASTM; 2006

[12] Schorr M. Corrosion and electrochemical behavior of austenitic stainless steels in phosphoric acid [MSc thesis]. Israel: Technion-Israel Institute of Technology; 1971

[13] Schorr M, Yahalom J. The significance of the energy of activation for the dissolution reaction of metal in acids. Corrosion Science. 1972;**12**:867

[14] Alon A, Schorr M, Yahalom I. Influence of halide in phosphoric acid on the corrosion behavior of stainless steel. Corrosion. 1975;**31**:319

[15] Kesten M, Peller HC. Über den einflub von adsorbierten anionenauf die magnetische umwandlung von Ni-Mo-legierungen. Corrosion Science. 1970;**10**: 401

[16] Riggs OL, Ray MH. Temperature coefficient of corrosion inhibition. Corrosion. 1967;**23**:125

[17] Smith AJ, Hatfield JD. Free corrosion of pure iron by phosphoric acid. Corrosion. 1970;**26**:246

[18] Hoar TP. The production and breakdown of the passivity of metals. Corrosion Science. 1967;7:341

[19] Quraishi MA, Sardar R. Corrosion inhibition of mild steel in acid solutions by some aromatic oxadiazoles. Materials Chemistry and Physics. 2002;**78**:425

[20] Quraishi MA, Sardar R. Hector bases–A new class of heterocyclic corrosion inhibitors for mild steel in acid solutions. Journal of Applied Electrochemistry. 2003;**33**:1163

[21] Moore WJ. Physical Chemistry. London: Longmans; 1998

[22] Cotrell AH. Theoretical Structural Metallurgy. Vol. 188. London: Edward Arnold; 1962. pp. 36-41

[23] Edeleanu C. In: Shrier LL, Jarman RA, Burstein GT, editors. Corrosion. London, Oxford: Butterworth-Heinemann; 1994

[24] Bockris JO'M, Subramanyan PK. Contributions to the electrochemical basis of the stability of metals. Corrosion Science. 1970;**10**:435

[25] Taleb A, Stafiej J, Badiali JP. Numerical simulation of metal corrosion with cluster formation. Transactions on Engineering Series. 2005;**48**:109

[26] Janik-Czachor M, Wood GC, Thompson GE. Assessment of the processes leading to pit nucleation. British Corrosion Journal. 1980;**15**:154

[27] Shewmon PC. Diffusion in Solids. New York: McGraw-Hill; 1963. p. 175

[28] Hull D. Introduction to Dislocations. Oxford: Pergamon Press; 1968. p. 21

Section 4

Biocorrosion

Chapter 9

Production of Hydroxyapatite on the Surface of Ti6Al7Nb Alloy as Compared to Ti6Al4V Alloy

Elinor Nahum, Svetlana Lugovskoy and Alex Lugovskoy

Abstract

Ti6Al4V is very commonly used for the production of dental implants. Titanium alloys whose mechanical and corrosion properties are equal or better than those of Ti6Al4V might present interest as plausible future materials, too. Ti6Al7Nb alloy was tested and compared to Ti6Al4V in this work. Samples of both alloys were oxidized in a water solution containing calcium acetate ($Ca(CH_3COO)_2$) and calcium glycerophosphate ($Ca(PO_4CH(CH_2OH)_2)$) by Plasma Electrolytic Oxidation (PEO) for 20 min. After that, the samples were hydrothermally treated (HTT) in water (pH = 7) and in potassium hydroxide (KOH) solution (pH = 11) for 2 hours at 200°C in a pressurized reactor. The content and morphology of hydroxyapatite (HA) layers formed on the surface of both alloys after the PEO and subsequent HTT treatments were studied. The surface morphologies, elemental composition, and phase components were characterized by Scanning Electron Microscopy (SEM), Energy Dispersive Spectroscopy (EDS), and X-Ray Diffraction (XRD), respectively. The surface roughness was measured by Atomic Force Microscope (AFM), and thickness measurements were made by SEM and thickness gauge. Corrosion measurements were performed for the comparison of the corrosion behavior of the two alloys.

Keywords: Ti6Al4V, Ti6Al7Nb, plasma electrolytic oxidation (PEO), hydrothermal treatment (HTT), hydroxyapatite (HA)

1. Introduction

Titanium alloys are often used for the production of various tools or devices to be implanted into a human body: artificial joints, blood vessel prostheses, dental implants, and so on. Of the most popular titanium alloys in that field are Ti6Al4V (Titanium grade 5) and Ti6Al4V-ELI (Titanium grade 23), which both have relatively low Young moduli, compatible with that of the bone issues, good fatigue strength, and excellent corrosion resistance in physiological environments [1]. A layer containing mainly Titania (TiO_2) is formed spontaneously on the surface of Titanium alloys. Not only does this layer protect the alloy against corrosion, but it also favors their integration with living tissues, that is, *osseointegration* [2].

Other titanium alloys having suitable properties might present both theoretical and applied interest as the novel materials for medical device production. A Niobium-containing Ti6Al7Nb is one of such alloys. The corrosion behavior of

IntechOpen

Ti6Al7Nb in the simulated body fluid (SBF) was studied by Rajendran et al. and was found comparable or better than that of Ti6Al4V-ELI [3].

While the bioinertness (including corrosion stability) of titanium alloys is high enough, their readiness to osseointegration leaves much to be desired [1]. One of the plausible strategies allowing a considerable improvement in the osseointegration of titanium alloys is the production of a layer of Hyfroxyapatite (HA) [4] on their surface. Being a mineral constituent of the bone tissue, Hydroxyapatite is an ideal binder between the metal and the living body.

In this study, the surface modification aiming at the production of HA on the surface of Ti6Al4V and Ti6Al7Nb was made by using Plasma Electrolyte Oxidation (PEO), which is a simple technique for producing hard and rough coating having numerous micro-pores [5, 6]. Using that technique, the insertion into the coating of such elements as calcium and phosphorous may be performed by just adding them to the electrolyte in a suitable form. A PEO layer may also present a diffusional and sorption barrier to the release of metal ions into physiological liquids, thus improving the bioinertness of the core metal [7]. PEO coatings often have good adhesion to the metal even if the implant geometry is complex such as screw-shaped implants [8, 9]. PEO by itself does not cause the growth of HA crystals, rather a specimen needs an additional hydrothermal treatment (HT) [4].

The aim of this study is to compare the efficacy of the production of HA on the surfaces of Ti6Al4V and Ti6Al7Nb by PEO and the subsequent HT.

2. Experimental

Ti6Al4V samples of 40 mm $\times$ 20 mm $\times$ 1 mm size and Ti6Al7Nb samples of 40 mm $\times$ 20 mm $\times$ 3 mm size were cut by laser and grounded by 150, 360, 600, and 1000 grid silicon carbide (SiC) papers. The specimens were rinsed in distilled water and acetone in an ultrasonic cleaner for 5 min. The PEO was performed by the 50 Hz sinusoidal AC current in an electrolyte containing 0.25 M calcium acetate and 0.06 M calcium glycerophosphate in distilled water at the current density of 4A/ dm^2 for 20 min. The PEO process occurred in a water-cooled stainless-steel container serving as the counter electrode, equipped with a mechanical stirrer. After the completion of the PEO process, the specimens were washed in distilled water and dried on air. After that, the specimens were hydrothermally treated in distilled water or in a KOH solution at 200°C in a pressurized reactor for 2 hours. The pressure during the treatment was 13–15 bar.

The surface morphology and elemental composition were characterized by scanning microscope electron (SEM) TESCAN MAIA3 TriglavTM equipped with AZteq Oxford energy dispersive spectroscopy (EDS) analyzer. X-Ray diffraction (XRD) Rigaku, SmartLab X-RAY DIFRACTOMETER using Cu-Kα radiation (λ = 1.54 Å) in the range of 15–65° angles with a step 0.02° was used to characterize the phase components of the substrates and coating. The thickness of the coatings was measured by ElektroPhysik MiniTest 730 thickness gauge based on eddy current principle by an average of 10 measurements. Focused ion beam (FIB) technique FEI Helios NanoLab™ 600 DualBeam was used for the production of cross-sectional area on a specimen to be further characterized by SEM-EDS. Surface roughness of the samples was evaluated with atomic force microscope (AFM) Bruker's Dimension FastScan with ScanAsystTM using the contact mode.

The corrosion resistance was determined on an IVIUMnSTAT potentiostat by electrochemical polarization methods, namely Linear Polarization Resistance (LPR) and Tafel Slope Extrapolation (TSE) using a three-electrode cell, where an Ag|AgCl electrode served as the reference electrode, and a platinum wire was the counter electrode. All the corrosion tests were performed in Hank's solution [10] and a simulated saliva solution [11], whose chemical compositions are given in **Table 1**. The pH of the electrolytes was 7, and the temperature was maintained at 36.5°C.

3. Results and discussion

The surface of both titanium alloys after the PEO has the typical for that technique microstructure characterized by microscopic pores scattered randomly across the surface. Cracks seen on the surface are more pronounced for Ti6Al4V (**Figure 1**).

After the hydrothermal treatment, the surface has changed. If the HT is performed in distilled water at pH = 7, Ti6Al4V surface is characterized by grainy HA crystals on the surface and very small needle-like HA crystals inside the pores (**Figure 2a**). Unlike that, numerous HA platelets are observed both inside and outside the pores on the surface of Ti6Al7Nb (**Figure 2b**). If the HT is made at pH = 11, the surface of Ti6Al4V is covered by ununiform plates of significantly larger HA crystals inside and outside the pores (**Figure 2c**). Ti6Al7Nb surface

Hank's solution [10]		Saliva solution [11]	
	Composition, g/L	**Reagent**	**Composition, g/L**
$CaCl_2{\bullet}2H_2O$	0.185	$MgCl_2{\cdot}6H_2O$	0.059
$MgSO_4$	0.09767	KCl	0.625
KCl	0.4	KH_2PO_4	0.326
KH_2PO_4	0.06	K_2HPO_4	0.804
$NaHCO_3$	0.35	$CaCl_2{\cdot}2H_2O$	0.166
NaCl	8.0	$C_3H_8O_3$	2.00
Na_2HPO_4	0.04788	Sodium carboxymethyl cellulose	10.0
Glucose	1.0		

Table 1.
Chemical composition of Hank's and simulated saliva solutions.

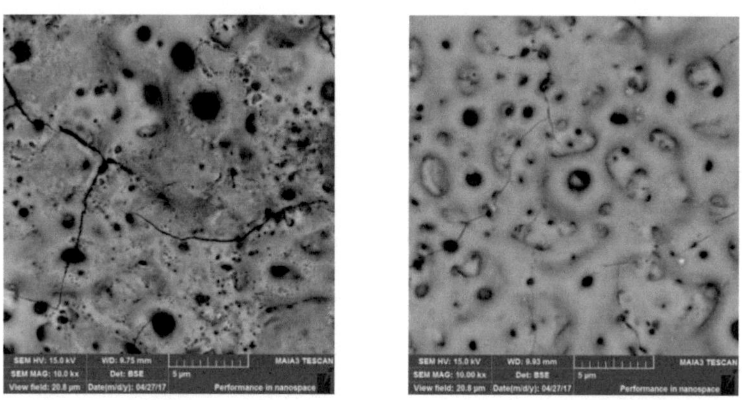

(a) (b)

Figure 1.
Back scattered electrons SEM images of the sample's surface after PEO, ×10,000: (a) Ti6Al4V and (b) Ti6Al7Nb.

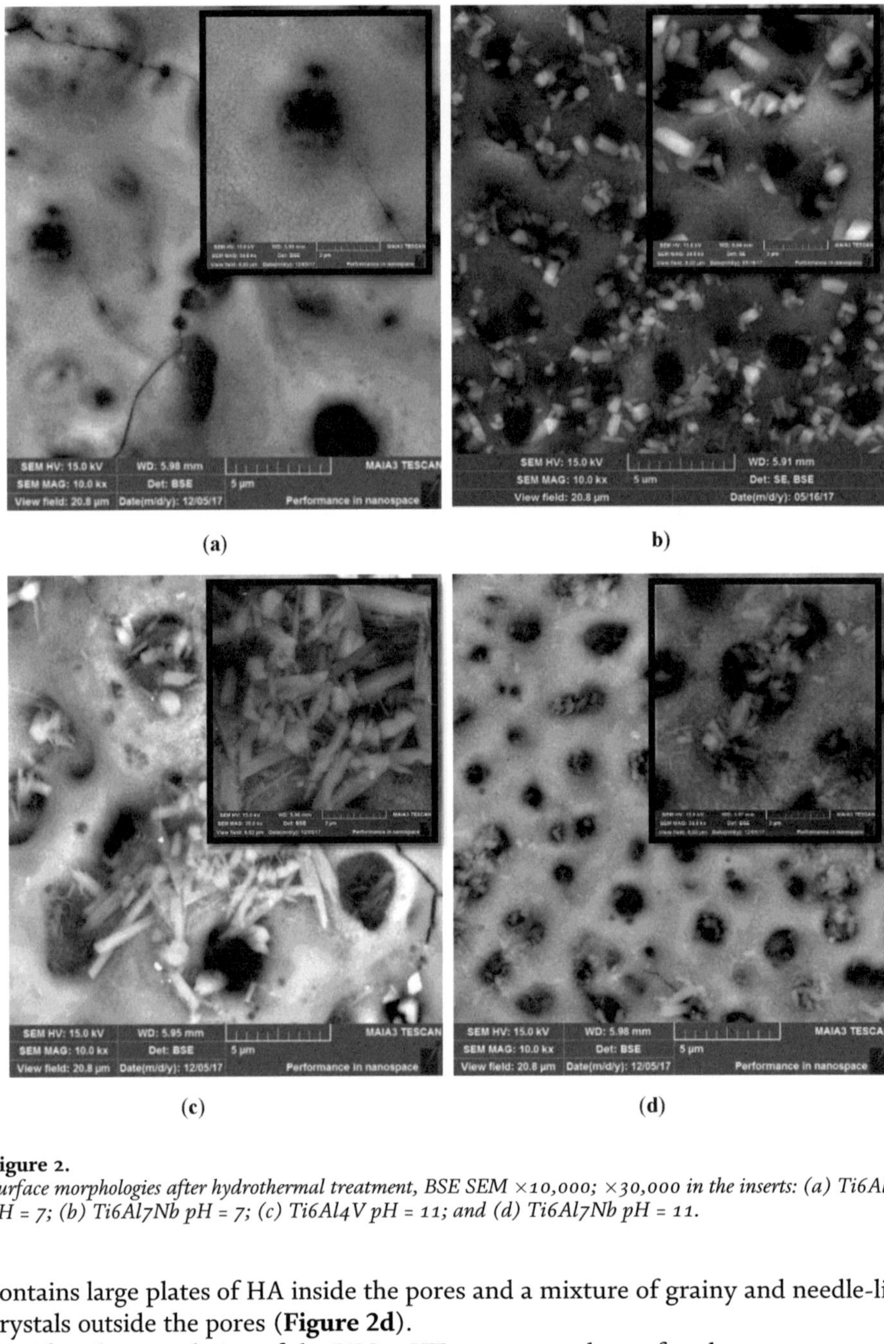

Figure 2.
Surface morphologies after hydrothermal treatment, BSE SEM ×10,000; ×30,000 in the inserts: (a) Ti6Al4V pH = 7; (b) Ti6Al7Nb pH = 7; (c) Ti6Al4V pH = 11; and (d) Ti6Al7Nb pH = 11.

contains large plates of HA inside the pores and a mixture of grainy and needle-like crystals outside the pores (**Figure 2d**).

After the completion of the PEO + HT treatment, the surface layers were partially ablated by FIB, so that the 'cross-sectional' structure could be seen (**Figure 3**). As is seen in **Figure 3**, there is an approximately 1 µm porous PEO oxide layer on the surface of the alloys. The oxide layer has partially amorphous and partially fine crystalline structure (region 'b' in the image); above that, an approximately 1 µm hydroxyapatite layer (region 'c' in the image) consisting of larger crystallites is present. The thicknesses of the oxide and hydroxyapatite layers may vary from one specimen to another, while their structure remains the same.

The elemental compositions of the surfaces obtained by EDS are given in **Table 2**. The presented chemical composition is the average of three-point mode analysis, and standard deviations are displayed. The stoichiometric Ca/P ratio for

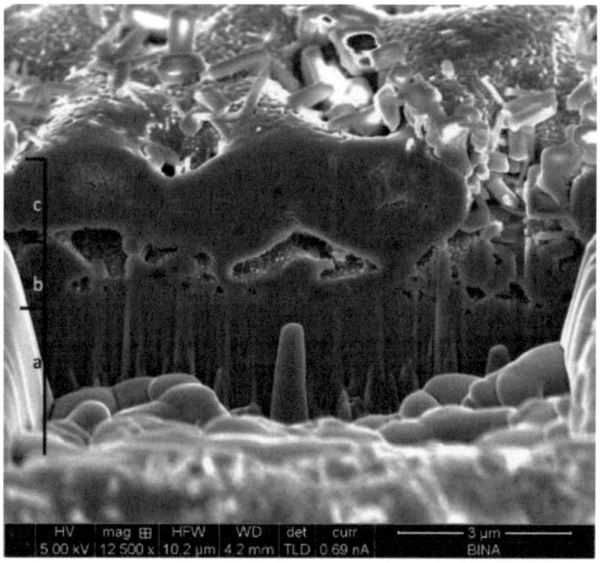

Figure 3.
A typical FIB-ablated cross-sectional structure of the surface: (a) titanium alloy substrate, (b) PEO porous layer, and (c) hydroxyapatite layer.

	Ti	Al	Nb or V	O	Ca	P	Ca/P
Ti6Al4V, PEO	15.1 ± 0.1	1.6 ± 0.0	0.6 ± 0.0	63.3 ± 0.2	8.6 ± 0.0	5.7 ± 0.0	1.51
Ti6Al7Nb, PEO	10.5 ± 0.5	1.2 ± 0.0	0.4 ± 0.0	58.1 ± 0.6	7.5 ± 0.1	4.6 ± 0.0	1.63
Ti6Al4V, HT pH = 7	17.7 ± 1.9	1.9 ± 0.0	3.0 ± 0.2	64.8 ± 0.6	7.2 ± 0.4	4.4 ± 0.2	1.64
Ti6Al7Nb, HT pH = 7	17.9 ± 0.3	2.2 ± 0.1	0.7 ± 0.1	69.8 ± 0.0	5.0 ± 0.3	4.6 ± 0.1	1.09
Ti6Al4V, HT pH = 11	15.5 ± 4.6	1.6 ± 0.7	0.4 ± 0.1	59.9 ± 7.7	11.6 ± 8.5	5.8 ± 3.4	2.00
Ti6Al7Nb, HT pH = 11	14.5 ± 3.4	2.1 ± 0.6	0.6 ± 0.0	65.3 ± 5.3	6.1 ± 3.7	5.1 ± 1.9	1.20

Table 2.
Elemental composition (at%, EDS) of the surfaces after PEO and hydrothermal treatments.

hydroxyapatite is 1.67 that was not observed for any specimen, which means that the surfaces always contain mixtures of various calcium phosphates rather than the pure hydroxyapatite.

In order to determine the phase composition of the surfaces, XRD spectra were measured (**Figure 4** and **Table 3**). It can be seen from **Figure 4** and **Table 3** that for both alloys no detectable amount of HA is present after PEO. Rather, the surfaces are covered by the mixture of rutile and anatase. Additionally, the surface of Ti6Al4V contains a small amount of tricalcium phosphate $Ca_2(PO_4)_2$, which is not the case for Ti6Al7Nb. After the HT treatment, an HA phase is detected on the surface of both alloys.

The thickness of the coating was determined by two different methods (**Table 4**), namely by using a thickness gauge and measuring the FIB-ablated cross sections in SEM images. The measurements reveal higher coating thicknesses for Ti6Al4V than Ti6Al7Nb. Ti6Al4V coating shows also larger and more uneven thickness for Ti6Al4V than for Ti6Al7Nb.

3D AFM images of Ti6Al4V and Ti6Al7Nb after PEO and hydrothermal treatments are shown in **Figure 5**. The area scanned was 5 μm × 5 μm, and three sites were scanned for each specimen. The values of average roughness (R_a) for all the

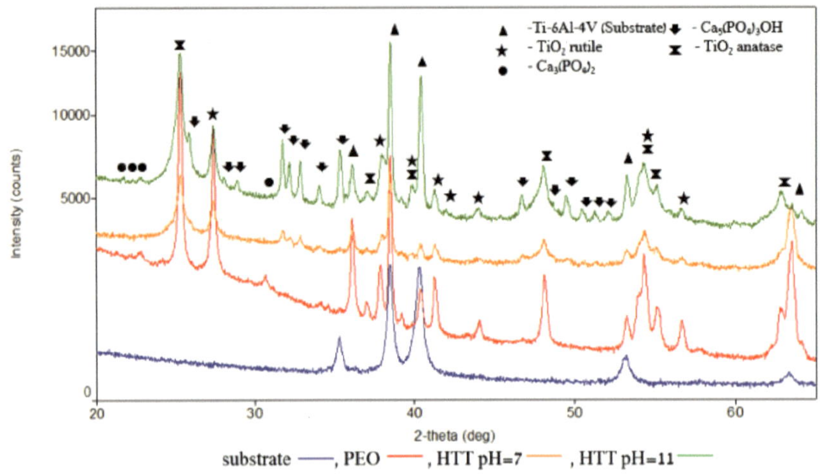

(a)

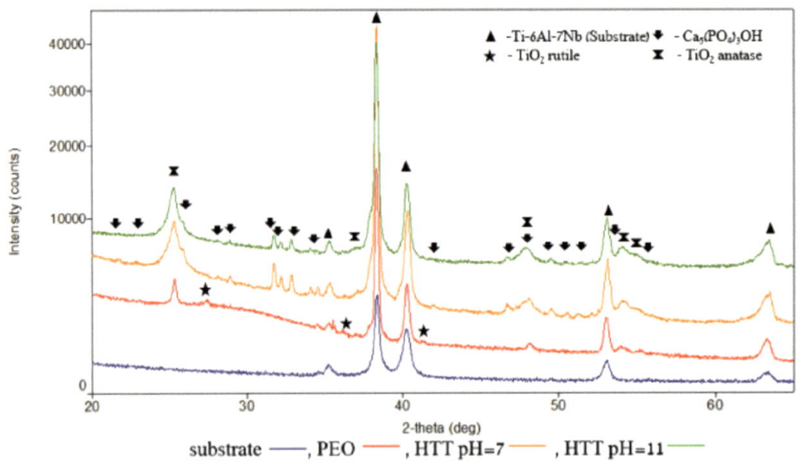

(b)

Figure 4.
XRD spectra acquired from the coatings: (a) Ti-6Al-4 V and (b) Ti-6Al-7Nb.

Treatment	Ti6Al7Nb	Ti6Al4V
PEO	TiO_2-rutile, anatase Amorphous phase	TiO_2-rutile, anatase $Ca_3(PO_4)_2$ Amorphous phase
HTT pH = 7	TiO_2-anatase HA	TiO_2-rutile, anatase HA
HTT pH = 11	TiO_2-anatase HA	TiO_2-rutile, anatase HA

Table 3.
Phase composition of coating after PEO and hydrothermal treatments.

specimens lie in the range of 50–250 nm (**Table 5**), and for Ti6Al7Nb, they are higher for all the treatments. As is seen from both AFM (**Figure 5**) and SEM (**Figure 2**) images, a more developed surface with plate-shaped HA crystals inside the pores and grainy crystals on the surface is formed on Ti6Al7Nb than on Ti6Al4V. The roughness range of 10 nm to 10 μm is favorable for the

Ti alloy	Treatment	Average coating thickness by SEM, μm	Average coating thickness by thickness gauge, μm
Ti6Al4V	PEO	6.4 ± 1.2	9.4 ± 1.0
	HTT pH = 7	8.5 ± 1.0	8.8 ± 1.1
	HTT pH = 11	7.0 ± 0.9	9.5 ± 1.2
Ti6Al7Nb	PEO	2.8 ± 0.4	2.8 ± 0.3
	HTT pH = 7	3.2 ± 0.1	2.4 ± 0.2
	HTT pH = 11	3.0 ± 0.6	2.6 ± 0.4

Table 4.
Thickness of the coatings on Ti alloys.

osseointegration because it is compatible with the sizes of small cells and large biomolecules [12, 13].

For the determination of corrosion parameters of the alloys, polarization curves of the specimens in Hank's solution and in artificial saliva were measured in the range of ±250 mV with respect to the OCP at the scan rate of 1 mV/s. Additionally, linear polarization measurements (LPRs) were performed in the narrower range of ±10 mV at the scan rate of 0.5 mV/s. The measured values of corrosion current densities and corrosion potentials are given in **Table 6**. For some cases (these are marked gray in **Table 6**), it was not possible to measure the corrosion parameters because the systems were too passive.

As is seen from **Table 6**, all the corrosion potentials that could be measured are significantly shifted to more noble values after the hydrothermal treatment, so that the alloy is effectively passivated. No essential difference was observed for the corrosion potential of the two alloys (at least, when those were measurable).

The values of corrosion current densities are scattered in a quite random manner and therefore are less informative. We assume that due to the relatively poor electrical conductivity of both liquids (the WC-CE resistance measured in the cell was ~10–20 kΩ for Hank's solution and for the artificial saliva, which is at least by the factor of 1000 higher than for such strong electrolytes as KCl), the precision of the polarization methods was not sufficient.

4. Conclusions

The morphologies, elemental and phase's composition, coating thickness, roughness, and corrosion behavior of Ti6Al4V and Ti6Al7Nb alloys after Plasma Electrolytic Oxidation and the subsequent hydrothermal treatment at various pHs were studied and compared. Hydroxyapatite-containing surfaces can be attained by the two-stage procedure, PEO and HTT, for both alloys.

Thicker, finer, and more uniform oxide layers are formed on the surface of Ti6Al4V than on Ti6Al7Nb for the same treatment parameters.

The most developed surface with plate-shaped HA crystals was obtained for Ti6Al7Nb after HTT in distilled water.

The corrosion potentials are significantly shifted to more noble values after the hydrothermal treatment, so that the alloy is effectively passivated. No essential difference was observed between the corrosion potential of the two alloys. It was

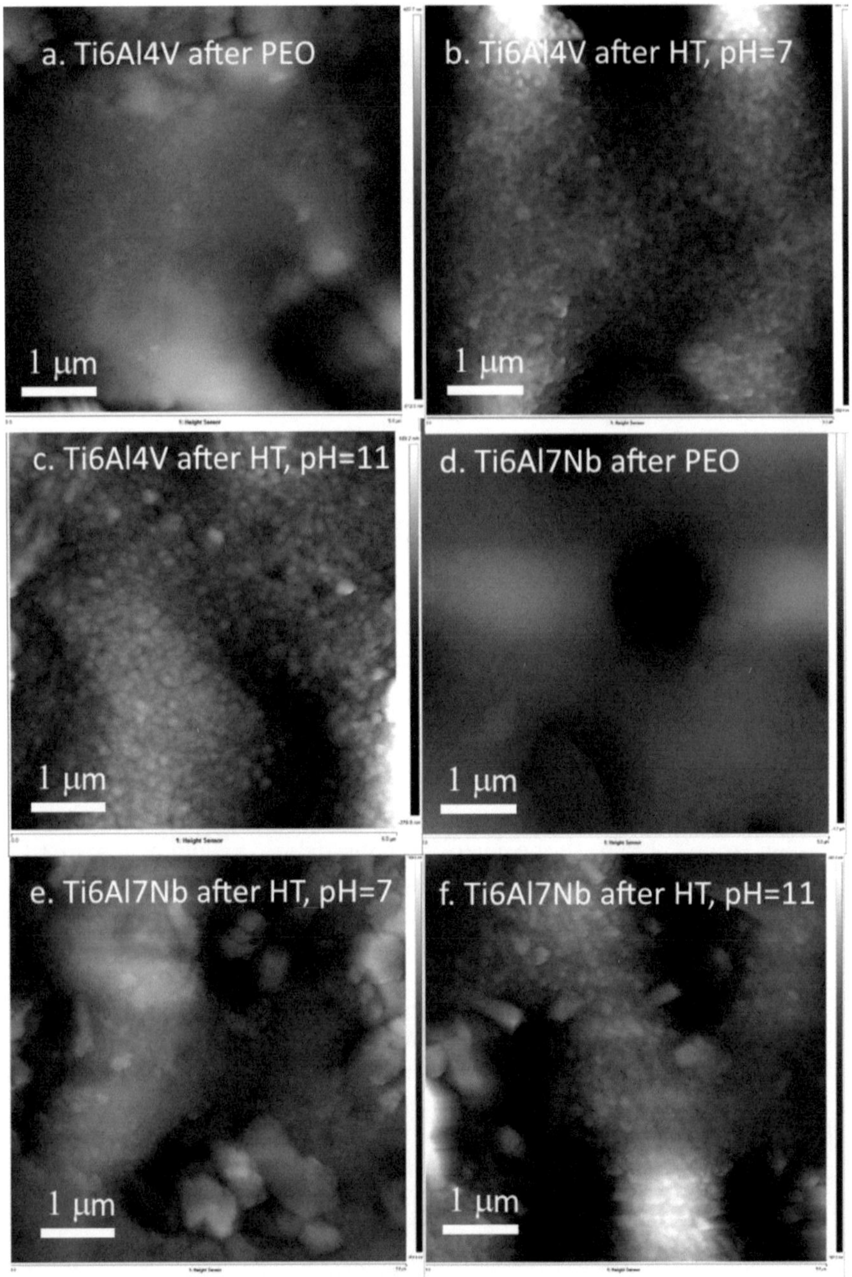

Figure 5.
3D AFM images: (a) Ti6Al4V after PEO; (b) Ti6Al4V after HTT pH = 7; (c) Ti6Al4V after HTT pH = 11; (d) Ti6Al7Nb after PEO; (e) Ti6Al7Nb after HTT pH = 7; (f) Ti6Al7Nb after HTT pH = 11.

	PEO	HT, pH = 7	HT, pH = 11
Ti6Al4V	83.1	115.7	55.4
Ti6Al7Nb	232.7	175.3	234.7

Table 5.
R_a *values (nm) for Ti6Al4V and Ti6Al7Nb surfaces after PEO and HT treatments.*

		E_{corr}, mV vs. Ag|AgCl	j_{corr}, A/cm^2, LPR	j_{corr}, A/cm^2, Tafel
Ti6Al4V, PEO	Hank's	−293	$3.76{\cdot}10^{-7}$	$2.31{\cdot}10^{-7}$
	Saliva	−303	$2.91{\cdot}10^{-8}$	$2.63{\cdot}10^{-8}$
Ti6Al7Nb, PEO	Hank's	−403	$2.66{\cdot}10^{-6}$	$1.32{\cdot}10^{-6}$
	Saliva	Passive		
Ti6Al4V, HT pH = 7	Hank's	−151	$5.16{\cdot}10^{-7}$	$2.83{\cdot}10^{-7}$
	Saliva	−156	$2.06{\cdot}10^{-7}$	$1.02{\cdot}10^{-7}$
Ti6Al7Nb, HT pH = 7	Hank's	Passive		
	Saliva	Passive		
Ti6Al4V, HT pH = 11	Hank's	−112	$1.38{\cdot}10^{-6}$	$1.23{\cdot}10^{-6}$
	Saliva	−127	$2.32{\cdot}10^{-7}$	$1.30{\cdot}10^{-7}$
Ti6Al7Nb, HT pH = 11	Hank's	Passive		
	Saliva	Passive		

Table 6.
Corrosion current density and corrosion potentials for Ti6Al4V and Ti6Al7Nb in Hank's and saliva solutions.

found that the polarization corrosion measurement is not precise enough for both alloys in Hank's solution and in the artificial saliva because of the poor conductivity of both liquids.

Author details

Elinor Nahum*, Svetlana Lugovskoy and Alex Lugovskoy
Department of Chemical Engineering, Ariel University, Ariel, Israel

*Address all correspondence to: elinorna@ariel.ac.il

IntechOpen

© 2020 The Author(s). Licensee IntechOpen. This chapter is distributed under the terms of the Creative Commons Attribution License (http://creativecommons.org/licenses/by/3.0), which permits unrestricted use, distribution, and reproduction in any medium, provided the original work is properly cited. CC BY

References

[1] Manivasagam G, Dhinasekaran D, Rajamanickam A. Biomedical implants: Corrosion and its prevention - A review. Recent Patents on Corrosion Science. 2010;**2**:40-54. DOI: 10.2174/1877610801002010040

[2] Jäger M, Jennissen HP, Dittrich F, Fischer A, Köhling HL. Antimicrobial and osseointegration properties of nanostructured titanium orthopaedic implans. Materials. 2017;**10**:1-28. DOI: 10.3390/ma10111302

[3] Tamilselvi S, Raman V, Rajendran N. Corrosion behaviour of Ti-6Al-7Nb and Ti-6Al-4V ELI alloys in the simulated body fluid solution by electrochemical impedance spectroscopy. Electrochimica Acta. 2006;**52**:839-846. DOI: 10.1016/j.electacta.2006.06.018

[4] Lugovskoy A, Lugovskoy S. Production of hydroxyapatite layers on the plasma electrolytically oxidized surface of titanium alloys. Materials Science and Engineering: C. 2014;**43**:527-532. DOI: 10.1016/j.msec.2014.07.030

[5] Lugovskoy A, Zinigrad M. Plasma electrolytic oxidation of valve metals. In: Mastai Y, editor. Advances in Materials Science and Engineering. London: InTech; 2013. pp. 85-102. DOI: 10.5772/54827

[6] Ibrahim MZ, Sarhan AAD, Yusuf F, Hamdi M. Biomedical materials and techniques to improve the tribological, mechanical and biomedical properties of orthopedic implants – A review article. Journal of Alloys and Compounds. 2017;**714**:636-667. DOI: 10.1016/j.jallcom.2017.04.231

[7] Mohedano M, Matykina E, Arrabal R, Pardo A, Merino MC. Metal release from ceramic coatings for dental implants. Dental Materials. 2014;**30**:e28-e40. DOI: 10.1016/j.dental.2013.12.011

[8] Zhu X, Chen J, Scheideler L, Reichl R, Geis-Gerstorfer J. Effects of topography and composition of titanium surface oxides on osteoblast responses. Biomaterials. 2004;**25**:4087-4103. DOI: 10.1016/j.biomaterials.2003.11.011

[9] Esen Z, Öcal EB. Surface characteristics and in-vitro behavior of chemically treated bulk Ti6Al7Nb alloys. Surface and Coatings Technology. 2017;**309**:829-839. DOI: 10.1016/j.surfcoat.2016.10.078

[10] Balanced Hank's Solution. Available from: https://www.sigmaaldrich.com/content/dam/sigma-aldrich/docs/Sigma/Formulation/d5796for.pdf

[11] Simulated Saliva Solution. Available from: https://www.pickeringlabs.com/wp-content/uploads/sds/SDS 1700 0305 Artificial Saliva for Medical and Dental Research.pdf

[12] Dilea M, Mazare A, Ionita D, Demetrescu I. Comparison between corrosion behaviour of implant alloys Ti6Al7Nb and Ti6Al4Zr in artificial saliva. Materials and Corrosion. 2013;**64**:493-499. DOI: 10.1002/maco.201206526

[13] Elias CN, Meirelles L. Improving osseointegration of dental implants. Expert Review of Medical Devices. 2010;**7**:241-256. DOI: 10.1586/erd.09.74

Chapter 10

Biotribology of Mechanically and Laser Marked Biomaterial

Marcelo de Matos Macedo, Vikas Verma,
Jorge Humberto Luna-Domínguez and Ronaldo Câmara Cozza

Abstract

The purpose of present work is to study the biotribological behavior of a mechanically and laser marked biomaterial. Sliding wear tests were conducted on ASTM F139 austenitic stainless-steel specimen, with polypropylene and AISI 316 L austenitic stainless-steel balls, as counterbodies. During wear experiments, a liquid chemical composition was continuously fed between the specimen and the ball. The coefficient of friction acting on the tribological system "specimen – liquid chemical composition – ball" and the wear volume of the wear craters were calculated, and results were analyzed. The results have shown that the biotribological behavior of ASTM F139 austenitic stainless steel was influenced by mechanical or laser marking process, and its wear resistance was dependent on the kind of counterbody.

Keywords: biomaterial, austenitic stainless steel, laser treatment, wear resistance, wear volume, coefficient of friction

1. Introduction

The "ball-cratering wear test" has gained large acceptance at universities and research centers as it is widely used in studies focusing on the wear behavior of different materials [1–20]. **Figure 1** presents a schematic diagram of the principle of wear test, where a rotating ball is forced against the tested specimen and liquid solution supplied between the specimen and the ball during the experiments.

The aim of the "ball-cratering wear test" is to generate "wear craters" on the surface of the specimen. **Figure 2** presents an image of such crater, together with an indication of the crater diameter (d) (**Figure 2a**) and the wear volume (V) (**Figure 2b** [21]). The wear volume is determined as a function of "d," using Eq. (1) [22], where R is the radius of the ball.

$$V \cong \frac{\pi d^4}{64R} \text{ for } d << R \quad (1)$$

In other line of research, the concept of "biotribology" has gained important spotlight in the area, including research works addressing the biotribological behavior of materials [23–29] used in the manufacturing of human body elements. Consequently, different laboratory techniques have been employed to reproduce conditions where there are friction and consequent wear of parts of the human mechanical structure with relative movement.

 IntechOpen

However, wear tests conducted under the "ball-cratering" technique present advantages in relation to other types of tribological procedures, as it favors the desired analysis of the tribological behavior. Therefore, considering the need of tribological characterization of biomaterials and the capacity that the "ball-cratering wear test" method presents to this goal, the purpose of this work is to study the biotribological behavior of mechanically and laser-marked ASTM F139 austenitic stainless-steel biomaterial.

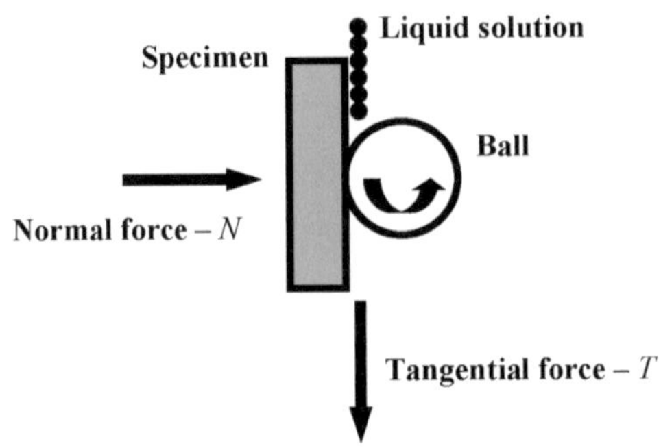

Figure 1.
"Ball-cratering wear test": representative figure of its operating principle.

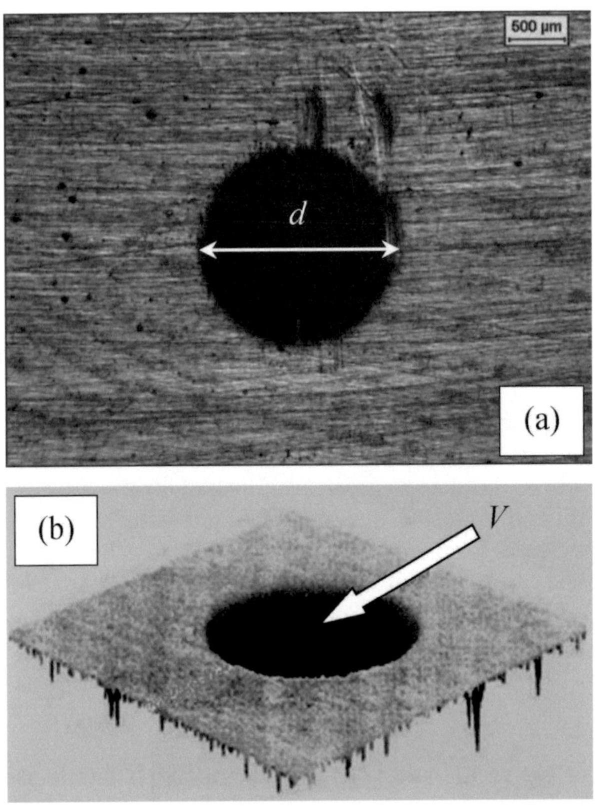

Figure 2.
Images of wear craters: (a) diameter – d and (b) wear volume – V [21].

2. Experimental details

2.1 Ball-cratering wear test equipment

Equipment with free-ball mechanical configuration (**Figure 3a**) was used for the sliding wear tests. Two load cells were used in the tribometer: one load cell to control the "normal force – *N*" applied on the specimen (**Figure 3b**) and the other load cell to measure the "tangential force – *T*" developed during the experiments (**Figure 3c**).

"Normal" and "tangential" force load cells have a maximum capacity of 50 N and an accuracy of 0.001 N. The values of *N* and *T* were registered by a data acquisition system, in real time, during the sliding wear tests.

2.2 Materials

The tested specimen was an ASTM F139 austenitic stainless-steel biomaterial, marked mechanically and with a nanosecond Q-switched Nd: YAG laser. Its chemical composition is presented in **Table 1**.

Balls of polypropylene and AISI 316 L austenitic stainless steel, with diameter of $D = 25.4$ mm ($D = 1''$ – standard size), were adopted as counterbodies.

To simulate the fluid present in the human body, a chemical liquid solution of PBS – Phosphate Buffered Solution – was inserted between the specimen and the ball. It was composed by the materials mentioned in **Table 2**.

Table 3 shows the hardness (*H*) of the materials used in this work (specimen and test balls).

2.3 Ball-cratering wear tests and data acquisition

Table 4 presents the test conditions defined for the sliding wear experiments conducted in this research.

Following values of normal force (*N*) for the sliding wear experiments: $N_{PP} = 0.05$ N and $N_{316L} = 0.40$ N were defined as a function of density (ρ) of the ball material – polypropylene $\Rightarrow \rho_{PP} = 0.91$ g/cm^3 and AISI 316 L austenitic stainless steel $\Rightarrow \rho_{316L} = 8$ g/cm^3. The rotational speed (n) of ball was 75 rpm. For $n = 75$ rpm and $D = 25.4$ mm, the tangential sliding velocity (v) of the ball is 0.1 m/s. Wear tests were conducted under a test time (t) of 10 min. With 0.1 m/s tangential sliding velocity and 10 min (600 s) test time, a sliding distance (S) of 60 m was calculated between the ball and the specimen.

All experiments were conducted without interruption, and the chemical liquid solution of PBS – Phosphate Buffered Solution – was fed between the specimen and the ball during the tests, under a frequency of 1 drop/10 s. Both the normal force (*N*) and the tangential force (*T*) were monitored and registered constantly. Finalizing the sliding wear tests, the diameters (d) of the wear craters were measured by optical microscopy, and their surfaces were analyzed by scanning electron microscopy.

Finally, the wear volume (V) was calculated by Eq. (1), and the coefficient of friction (μ) was determined using Eq. (2):

$$\mu = \frac{T}{N} \tag{2}$$

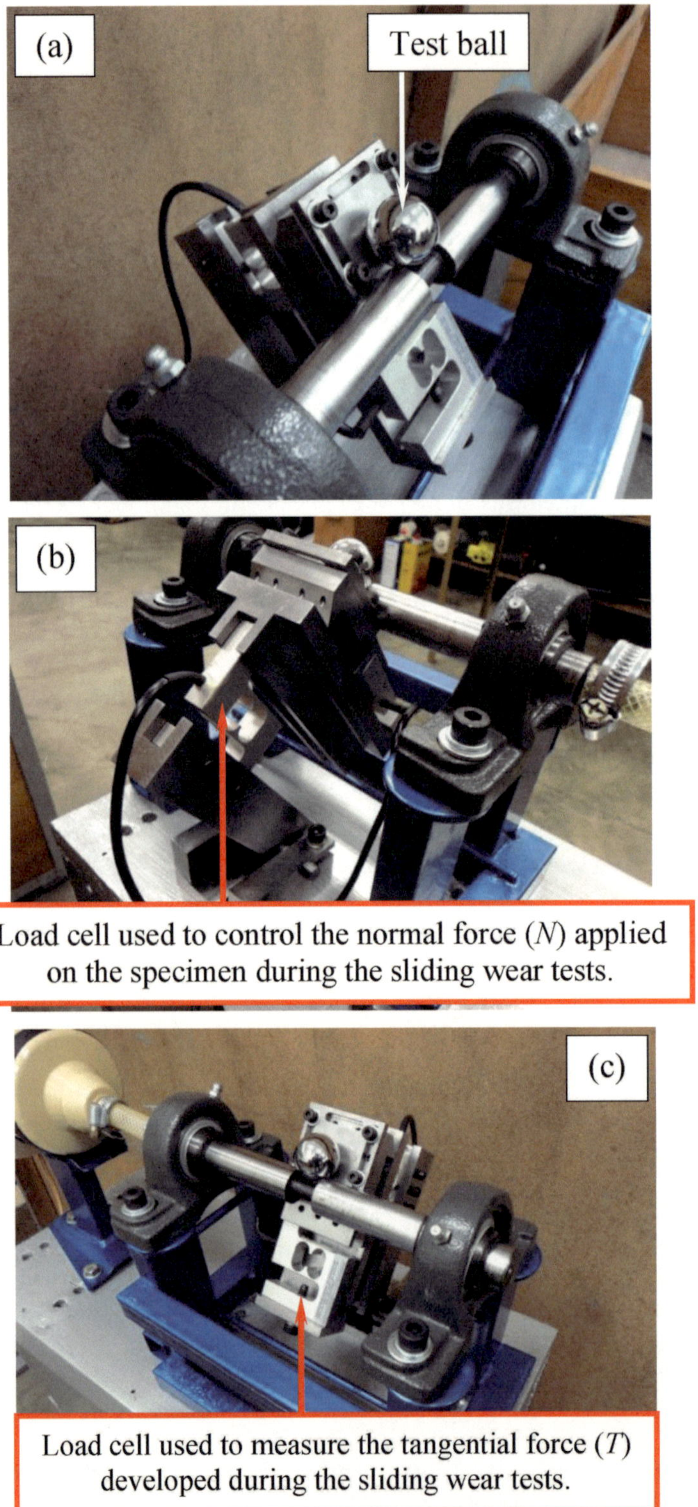

Figure 3.
(a) "Ball-cratering" wear test equipment with "free-ball" mechanical configuration used for the sliding wear tests: (b) load cell mounted to control the normal force and (c) load cell positioned to measure the tangential force during the experiments.

Chemical element	% (in weight)
C	0.023
Si	0.78
Mn	2.09
P	0.026
S	0.0003
Cr	18.32
Mo	2.59
Ni	14.33
Fe	Balance

Table 1.
Chemical composition of ASTM F139 austenitic stainless-steel biomaterial – in percentage weight.

Chemical element	(g/l)
NaCl	8
KCl	0.2
Na_2HPO_4	1.15
KH_2PO_4	0.2

Table 2.
Chemical composition of the PBS – phosphate buffered solution – in g/l.

	Material	Hardness – *H*
Specimen	ASTM F139 austenitic stainless steel	180 HV
Test ball	Polypropylene	55 – Shore D
	AISI 316 L austenitic stainless steel	318 HV

Table 3.
Hardness of the materials used in this work.

Normal force – N_{PP}	Ball of polypropylene	0.05 N
Normal force – N_{316L}	Ball of AISI 316 L austenitic stainless steel	0.40 N
Test ball rotational speed – *n*		75 rpm
Tangential sliding velocity – *v*		0.1 m/s
Test time – *t*		10 min
Sliding distance – *S*		60 m

Table 4.
Test parameters for the ball-cratering wear tests under conditions of sliding wear.

3. Results and discussion

3.1 Scanning electron microscopy

Figure 4 shows a scanning electron micrograph of the surface of a wear crater generated during the sliding wear tests.

Occurrence of grooves, due to sliding movement between the ball and the specimen, was observed in the scanning electron micrograph. The result presented in **Figure 4** is in qualitative agreement with the literature [30], where it is reported that the action of grooves on the surface of a material is characterized as a common tribological behavior of two metallic materials under relative movement.

3.2 Wear volume behavior

Figure 5 presents the behavior of the specimen in terms of wear volume (*V*) for the following conditions: mechanically and laser-marked specimen and different types of balls (counterbodies).

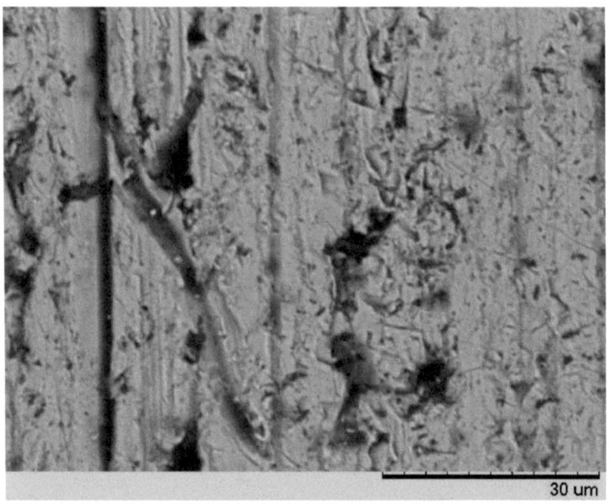

Figure 4.
Scanning electron micrograph of the surface of a wear crater generated during the sliding wear tests.

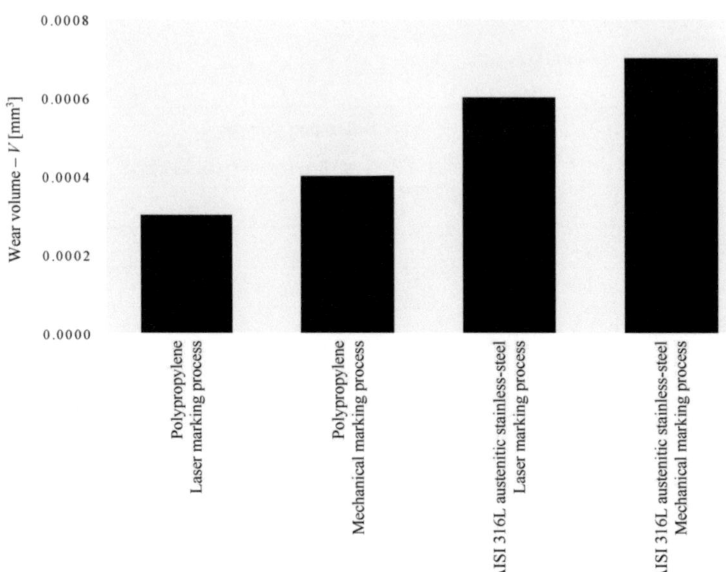

Figure 5.
Wear volume (V) behavior as a function of the type of marking process ("mechanical" or "laser") and type of counterbody (ball of polypropylene or ball of AISI 316 L austenitic stainless steel). Maximum standard deviation reported: ±7 × 10^{-4} mm^{3}.

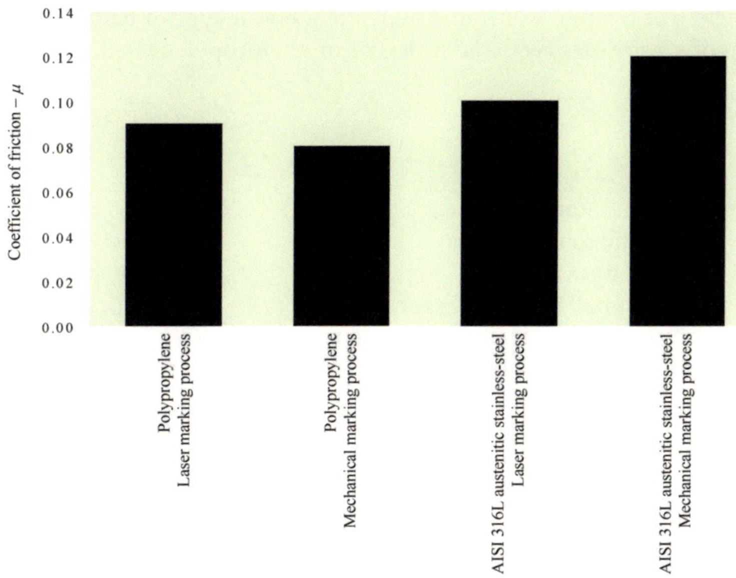

Figure 6.
Coefficient of friction behavior (μ) as a function of the type of marking process ("mechanical" or "laser") and type of counterbody (ball of polypropylene or ball of AISI 316 L austenitic stainless steel). Maximum standard deviation reported: ±0.03.

In addition, **Figure 5** shows a decrease in wear volume under laser marking process for both types of counterbodies. Decrease in wear volume is related to increase in local hardness of the specimen. Increase of hardness can be attributed to the action of laser on specimen surface. In relation to specimen marked mechanically, the possible increase of local hardness could have occurred due to local surface hardening.

However, the increase of local hardness caused by laser marking is higher than the local hardness caused by mechanical marking, justifying the results presented in **Figure 5**.

3.3 Coefficient of friction behavior

Figure 6 shows the behavior of the coefficient of friction (μ) for the conditions, which the specimen is marked mechanically and marked with laser, and for the different types of balls – counterbodies.

In the present tribological conditions, coefficient of friction was found lower for the wear tests conducted against polypropylene ball than AISI 316 L austenitic stainless-steel ball counterbodies.

4. Conclusions

The following conclusions can be drawn from the results obtained in this research, regarding to tribological behavior of ASTM F139 austenitic stainless steel:

- tribological behavior was influenced by the type of the marking process – "mechanical" or "laser" – applied for the investigated biomaterial;
- wear volume was found to be dependent on the normal force acting on the specimen, that is, they were dependent on the type of counterbody – ball of polypropylene or ball of AISI 316 L austenitic stainless steel; and

- coefficient of friction was found dependent on the type of ball; the lower values of μ were observed under the use of polypropylene ball.

Nomenclature

d	diameter of the wear crater (mm)
D	diameter of the test ball (mm)
H	hardness (HV)
n	test ball rotational speed (rpm)
N	normal force (applied on the specimen) (N)
R	radius of the test ball (mm)
S	sliding distance (m)
t	test time (min)
T	tangential force (developed during the wear tests) (N)
v	tangential sliding velocity of the test ball (m/s)
V	wear volume of the wear crater (mm^3)

Greek letters

μ	coefficient of friction
ρ	density (g/cm^3)

Author details

Marcelo de Matos Macedo[1], Vikas Verma[2], Jorge Humberto Luna-Domínguez[3] and Ronaldo Câmara Cozza[1,4]*

1 Department of Mechanical Manufacturing, Technology Faculty – FATEC-Mauá, CEETEPS – State Center of Technological Education "Paula Souza", Mauá, SP, Brazil

2 Thermochemistry of Materials SRC, National University of Science and Technology (NUST) MISiS, Moscow, Russia

3 Facultad de Odontología, Universidad Autónoma de Tamaulipas, Tamaulipas, Mexico

4 Department of Mechanical Engineering, University Center FEI – Educational Foundation of Ignatius "Padre Sabóia de Medeiros", SP, Brazil

*Address all correspondence to: rcamara@fei.edu.br

IntechOpen

© 2020 The Author(s). Licensee IntechOpen. This chapter is distributed under the terms of the Creative Commons Attribution License (http://creativecommons.org/licenses/by/3.0), which permits unrestricted use, distribution, and reproduction in any medium, provided the original work is properly cited.

References

[1] Cozza RC. A study on friction coefficient and wear coefficient of coated systems submitted to micro-scale abrasion tests. Surface & Coatings Technology. 2013;**215**:224-233. DOI: 10.1016/j.surfcoat.2012.06.088

[2] Cozza RC, Rodrigues LC, Schön CG. Analysis of the micro-abrasive wear behavior of an iron aluminide alloy under ambient and high-temperature conditions. Wear. 2015;**330-331**:250-260. DOI: 10.1016/j.wear.2015.02.021

[3] Cozza RC, JTSL W, Schön CG. Influence of abrasive wear modes on the coefficient of friction of thin films. Tecnologia em Metalurgia, Materiais e Mineração. 2018;**15**(4):504-509. DOI: 10.4322/2176-1523.20181468

[4] Umemura MT, Jiménez LBV, Pinedo CE, Cozza RC, Tschiptschin AP. Assessment of tribological properties of plasma nitrided 410S ferritic-martensitic stainless steels. Wear. 2019;**426-427**:49-58. DOI: 10.1016/j.wear.2018.12.092

[5] Cozza RC. Thin films: Study of the influence of the micro-abrasive wear modes on the volume of wear and coefficient of friction. In: Friction, Lubrication and Wear. 1st ed. London–UK: IntechOpen; 2019. DOI: 10.5772/intechopen.86459

[6] Cozza RC. Effect of pressure on abrasive wear mode transitions in micro-abrasive wear tests of WC-Co P20. Tribology International. 2013;**57**:266-271. DOI: 10.1016/j.triboint.2012.06.028

[7] Cozza RC. Effect of sliding distance on abrasive wear modes transition. Journal of Materials Research and Technology. 2015;**4**(2):144-150. DOI: 10.1016/j.jmrt.2014.10.007

[8] Wilcken JTSL, Silva FA, Cozza RC, Schön CG. Influence of abrasive wear modes on the coefficient of friction of thin films. In: Proceedings of the "ICAP 2014 – 2nd International Conference on Abrasive Processes". Cambridge – UK: The University of Cambridge; September 8-10, 2014

[9] Cozza RC. Study of the steady-state of wear in micro-abrasive wear tests by rotative ball conducted on specimen of WC-Co P20 and M2 tool-steel. Revista Matéria. 2018;**23**(1):e-11986. DOI: 10.1590/s1517-707620170001.0322

[10] da Silva WM, Binder R, de Mello JDB. Abrasive wear of steam-treated sintered iron. Wear. 2005;**258**:166-177. DOI: 10.1016/j.wear.2004.09.042

[11] Trezona RI, Allsopp DN, Hutchings IM. Transitions between two-body and three-body abrasive wear: Influence of test conditions in the microscale abrasive wear test. Wear. 1999;**225-229**:205-214. DOI: 10.1016/S0043-1648(98)00358-5

[12] Adachi K, Hutchings IM. Wear-mode mapping for the micro-scale abrasion test. Wear. 2003;**255**:23-29. DOI: 10.1016/S0043-1648(03)00073-5

[13] Adachi K, Hutchings IM. Sensitivity of wear rates in the micro-scale abrasion test to test conditions and material hardness. Wear. 2005;**258**:318-321. DOI: 10.1016/j.wear.2004.02.016

[14] Cozza RC, de Mello JDB, Tanaka DK, Souza RM. Relationship between test severity and wear mode transition in micro-abrasive wear tests. Wear. 2007;**263**:111-116. DOI: 10.1016/j.wear.2007.01.099

[15] Cozza RC, Tanaka DK, Souza RM. Micro-abrasive wear of DC and pulsed DC titanium nitride thin films with different levels of film residual stresses. Surface and Coatings Technology.

2006;**201**:4242-4246. DOI: 10.1016/j.surfcoat.2006.08.044

[16] Bose K, Wood RJK. Optimun tests conditions for attaining uniform rolling abrasion in ball cratering tests on hard coatings. Wear. 2005;**258**:322-332. DOI: 10.1016/j.wear.2004.09.018

[17] Mergler YJ, Huis in 't Veld AJ. Micro-abrasive wear of semi-crystalline polymers. Tribology and Interface Engineering Series. 2003;**41**:165-173. DOI: 10.1016/S0167-8922(03)80129-3

[18] Batista JCA, Matthews A, Godoy C. Micro-abrasive wear of PVD duplex and single-layered coatings. Surface and Coatings Technology. 2001;**142-144**:1137-1143. DOI: 10.1016/S0257-8972(01)01189-6

[19] Batista JCA, Godoy C, Matthews A. Micro-scale abrasive wear testing of duplex and non-duplex (single-layered) PVD (Ti,Al)N, TiN and Cr-N coatings. Tribology International. 2002;**35**:363-372. DOI: 10.1016/S0301-679X(02)00017-8

[20] Batista JCA, Joseph MC, Godoy C, Matthews A. Micro-abrasion wear testing of PVD TiN coatings on untreated and plasma nitrided AISI H13 steel. Wear. 2002;**249**:971-979. DOI: 10.1016/S0043-1648(01)00833-X

[21] da Silva WM. Effect of pressing pressure and iron powder size on the micro-abrasion of steam-oxidized sintered iron [M.Sc. Dissertation]. Uberlândia–MG, Brazil: Federal University of Uberlândia; 2003. p. 98

[22] Rutherford KL, Hutchings IM. Theory and application of a micro-scale abrasive wear test. Journal of Testing and Evaluation. 1997;**25**(2):250-260. DOI: 10.1520/JTE11487J

[23] Niinomi M, Nakai M, Hieda J. Development of new metallic alloys for biomedical applications. Acta Biomaterialia. 2012;**8**:3888-3903. DOI: 10.1016/j.actbio.2012.06.037

[24] Talha M, Behera CK, Sinha OP. A review on nickel-free nitrogen containing austenitic stainless steels for biomedical applications. Materials Science and Engineering C. 2013;**33**:3563-3575. DOI: 10.1016/j.msec.2013.06.002

[25] Gurappa I. Development of appropriate thickness ceramic coatings on 316 L stainless steel for biomedical applications. Surface and Coatings Technology. 2002;**161**:70-78. DOI: 10.1016/S0257-8972(02)00380-8

[26] Shih CC, Shih CM, Su YY, LHJ S, Chang MS, Lin SJ. Effect of surface oxide properties on corrosion resistance of 316L stainless steel for biomedical applications. Corrosion Science. 2004;**46**:427-441. DOI: 10.1016/S0010-938X(03)00148-3

[27] Hosseinalipour SM, Ershad-Langroudi A, Hayati AN, Nabizade-Haghighi AM. Characterization of sol–gel coated 316L stainless steel for biomedical applications. Progress in Organic Coatings. 2010;**67**:371-374. DOI: 10.1016/j.porgcoat.2010.01.002

[28] Dewidar MM, Khalil KA, Lim JK. Processing and mechanical properties of porous 3 16L stainless steel for biomedical applications. Transactions of Nonferrous Metals Society of China. 2007;**17**(3):468-473. DOI: 10.1016/S1003-6326(07)60117-4

[29] Niinomi M. Recent metallic materials for biomedical applications. Metallurgical and Materials Transactions A. 2002;**33A**:477-486. DOI: 10.1007/s11661-002-0109-2

[30] Cozza RC. Third abrasive wear mode: Is it possible? Journal of Materials Research and Technology. 2014;**3**(2):191-193. DOI: 10.1016/j.jmrt.2014.03.010